"Janet Kellogg Ray combines tran[...]
gritty commitment to science. *Baby D[...]*
new possibilities for bridge-building between the truth of faith and the facts of science. If you are tired of clumsy science and combative religion, Ray is the conversation partner you have been looking for! This book is for anyone who seeks truth wherever truth may be found."

— Don McLaughlin
senior minister of North Atlanta Church of Christ

"It's often unwise to judge a book by its title, yet this is the rare case when you should do just this. *Baby Dinosaurs on the Ark?* is an intriguing title completely backed up with an even more intriguing book. Janet Kellogg Ray blends storytelling, biology, and biblical reflection to offer a very helpful, engaging, and important book. All pastors, parents, and young adults will find this book an essential resource in understanding faith and science and a way to faithfully embrace them both."

— Andrew Root
author of *Exploding Stars, Dead Dinosaurs, and Zombies: Youth Ministry in the Age of Science*

"My wife, like Janet Kellogg Ray, is a science teacher. Her students and colleagues know she is married to a pastor. Each year, like clockwork, a student or fellow teacher asks her about the intersection of science and faith. Their assumption is that her allegiances lie with either faith or science, that she couldn't hold them both appropriately. Science and faith are in a dance together, and Janet Kellogg Ray's *Baby Dinosaurs on the Ark?* helps those of us with questions about the interplay of faith and science articulate and understand our faith better. Here you will discover more of what God is up to in the world, how faith and science testify to one another, and even more so how they testify the beauty of our Creator."

— Sean Palmer
author of *Unarmed Empire: In Search of Beloved Community*

"Ray writes with candid humor, a pastoral spirit, and engaging, accessible science. This book deserves to be widely read, especially if you're not sure that evolution and robust faith can go together."

— Dennis R. Venema
professor of biology at Trinity Western University

"What a delight to read! With an engaging style and a keen mind, Ray navigates the landscape between the false binary that so many Christians face: reject science or reject God. A trustworthy guide, Ray explores the various positions with intellectual honesty and civility; rare is the author who can explain this complex topic in such a clear and compelling way. If you are looking for a resource that equips you both to embrace the findings of science and to embody a deep faith, this is the book for you."

— **Ken Cukrowski**
dean of the College of Biblical Studies at
Abilene Christian University

"Too much Christian opinion on science has been uninformed and unhelpful. In *Baby Dinosaurs on the Ark?* Dr. Ray gives us a down-to-earth yet thorough introduction for how science works and how necessary it is to shake off unhelpful and untrue assumptions about the Bible. If anyone asks why you accept the science of evolution as a Christian, feel free to simply pass them a copy of this book."

— **Jared Byas**
cohost of the podcast
The Bible for Normal People

"This is the most cleverly written and yet profound book I've read in some time. I love it! Ray makes complex and deep issues accessible. She answers questions about science and contemporary debates. I plan to give copies to friends trying to make sense of evolution and Christian faith."

— **Thomas Jay Oord**
author of *The Uncontrolling Love of God*

Baby Dinosaurs on the Ark?

The Bible and Modern Science and the Trouble of Making It All Fit

JANET KELLOGG RAY

William B. Eerdmans Publishing Company
Grand Rapids, Michigan

Wm. B. Eerdmans Publishing Co.
4035 Park East Court SE, Grand Rapids, Michigan 49546
www.eerdmans.com

Published 2021
Printed in the United States of America

27 26 25 24 23 22 21 1 2 3 4 5 6 7

ISBN 978-0-8028-7944-8

Library of Congress Cataloging-in-Publication Data

Names: Ray, Janet Kellogg, 1960– author.
Title: Baby dinosaurs on the ark? : the Bible and modern science and the trouble of making it all fit / Janet Kellogg Ray.
Description: Grand Rapids, Michigan : Wm. B. Eerdmans Publishing Co., [2021] | Includes bibliographical references and index. | Summary: "An exploration of creationism and the science of origins that shows how a literal reading of the Bible—particularly the book of Genesis—can lead to distortions of scientific reality"—Provided by publisher.
Identifiers: LCCN 2021010006 | ISBN 9780802879448
Subjects: LCSH: Bible and science. | Bible—Criticism, interpretation, etc. | Creationism. | Bible and evolution.
Classification: LCC BS650 .R39 2021 | DDC 261.5/5—dc23
LC record available at https://lccn.loc.gov/2021010006

To Mark,
for believing that I have something worth saying

Contents

Foreword

Can the Bible and science work together? Perhaps you have been told no. Maybe your science professor said that you can't be a scientist and believe the Bible. Maybe your church preached that you must defend your faith against science. Maybe you heard both at once! Unfortunately, that "no" has become more shrill as polarization grows. Increasingly, people define their own identity by their disagreement with the other side, making it hard to see any value or agreement with others. Misinformation on science abounds. Social media pulls us further apart.

The "no" has driven countless people away from God. Secular scientists have only to hear about baby dinosaurs on a boat with humans to decide that they can't take Christianity seriously. And Christians have only to hear scientists openly mock their faith to decide that science is against them, such as in 2019 when Neil DeGrasse Tyson tweeted on December 25, "On this day long ago, a child was born who, by age 30, would transform the world. Happy Birthday Isaac Newton." The next generation is hearing the "no" loud and clear. In 2018, half of churchgoing teenagers agreed that "the church seems to reject much of what science tells us about the world" (*Gen Z: The Culture, Beliefs and Motivations Shaping the Next Generation* [Barna, 2018]). With the many news stories in 2020 of evangelicals skeptical about public health science on Covid-19, I expect that percentage has only grown. Unfortunately, the polarization has eternal consequences—issues of science are among the main reasons young people are leaving the church (David Kinnaman, *You Lost Me* [Grand Rapids: Baker Books, 2016]).

For me, and for author Janet Kellogg Ray, the answer once was no, but today is yes. I grew up in an evangelical church, learning the Bible from cover to cover and growing in my commitment to Christ. At our church we believed the earth was young, because the only alternative was an atheistic version of evolution and the big bang. "Evolution" was definitely a bad word. Yet I loved science, and the same church encouraged me to pursue it. When I had questions in high school about evolution, it helped immensely that my parents didn't tell me the "no"—they actually came alongside me in my confusion and said "I don't know." When I chose astronomy for my career, I finally took the time to dig into the issue. I was helped immensely when I discovered books by Christian astronomers, geologists, and biologists—fellow believers who explained the scientific evidence without an atheistic bias. They introduced me to biblical scholars and theologians who became the key for me to realize how I could say yes without denying the authority of the Bible. In fact, they led me to a richer understanding of Scripture and a deeper walk with God.

If you are asking similar questions, wondering if there is any way that Christian faith and evidence-based science can work together, Janet Kellogg Ray is an able guide. She is a biology teacher and a Christ-follower who invites you to walk alongside her in her journey and provides an engaging overview of the views, evidence, and arguments on origins science. I pray that for you, too, the exploration of these questions will bring you to a richer, deeper faith in which you love God with all your heart, soul, *and* mind.

Deborah Haarsma
Astronomer, President of BioLogos
Grand Rapids, Michigan
January 2021

Acknowledgments

"You should write a book," he said.

"You really need to write a book."

"Here. Read this fortune cookie: 'You are a lover of words, someday you should write a book.'"

To my husband, Mark, for prodding me forward, for reading endless rough drafts, for the spontaneous "a publisher wants my book!" party, and for finding that fortuitous fortune cookie.

To my book study besties: Sheila Legan, Melinda Ballard, Darlene Wood, Karen Berryman, Dede Jackson, Cynthia Heaberlin, Jeanie Virden, Jo Pummill, and Jennifer Mynhier. You are a safe place for the hard questions and the incubator that births renewed vision and faith.

To Melinda Ballard, soul sister and cheerleader and the only person I know who preorders ten copies of a book, sight unseen.

To my parents, Kirby and Lou Ann Kellogg, my lifelong champions who model faithfulness to Scripture, love for the church, and commitment to family.

To my uncle, the late Dr. Lynn Mitchell, and my dear aunt Carol Mitchell, who always encouraged my preaching and teaching.

And finally, loads of thanks to the Wm. B. Eerdmans team for their encouragement and guidance and for patiently answering my endless questions: Laurel Draper, Michael Debowski, Alexis Cutler, Laura Bardolph Hubers, and Amy Kent. Thank you especially to Trevor Thompson for welcoming a new writer.

1

The Biology Professor Who Doesn't Believe in Science

I teach biology, but I don't believe in science.

The first day of each semester begins the same way—a short discussion about the philosophy of science: what is science? What makes something science? What is not science?

I believe in a lot of things, I explain, but I don't believe in science.

It's shocking, I know . . . the biology professor does not believe in science! First-day-of-class engaged expressions turn to cautious bewilderment.

"Is it too late to transfer out of this course?"

There are many things I believe, but science is not one of them. Instead, I *accept* science evidence. After all, a fact is true whether I believe it or not.

I include evolution concepts throughout the semester, but during the last few weeks of the course, we dive headlong into to the details. We discuss what evolution is and is not. I passionately and unequivocally declare, "Evolution theory says nothing about God or religion or any other world view, for that matter."

Evolution theory says nothing about God or religion or any other world view, for that matter.

Because I teach in a public university, I cannot overtly say, “and this is why I am a Christian who accepts evolution,” but I’d love to have the discussion.

I tell my students about the very loud voices on both ends of the spectrum who say religious faith and evolution can never coexist. Then, as clearly as I can, I make the point: this just isn’t so.

At least I *thought* I was making it clear.

Each class period ends with a quick “minute paper”—students write a brief response to a concept from the lecture on an index card. On the last day of the semester, this was the writing prompt: “Evolution is a tinkerer. Explain.”

In the stack of almost 150 cards, two cards immediately stood out. Most students write a sentence or two and are out the door and on their way. These two cards, however, were covered with writing. One student also filled the back.

“I’ve listened to all your lectures, but I can disprove it all.”

“Evolution as you describe it goes against what God says and what I know to be true.”

And there was Scripture. Lots and lots of Scripture, quotes and references too—it was impressive.

There was more: “Evolution doesn’t tinker. God does, and only Him.”

And this declaration: “I will not deny.” One student tried to soften the blow a bit: “I’m not trying to be rude, I thoroughly enjoyed your class, but I won’t answer ‘correctly.’”

I am not sure what the emotion was I felt as I read these two cards. I definitely had a lump in my throat. In the opinion of these two students, I, their professor, am asking them to deny God, and they aren’t going to do it.

They intend to stand up for Jesus, even if it means a bad grade on the notecard.

This is probably not fair, but I felt "put in my place." These students think I don't know anything about the Bible! Me—with all my Bible Bowl coaching and Bible for Credit classes and a lifetime of church!

Two were brave enough to write it down and turn it in; how many more thought the same but did not want to risk a bad grade? Puzzlingly, my overwhelming emotion was embarrassment, but why?

Then it hit me: it's me. I am the professor their parents and pastors warned them about. I'm the scary atheist professor in all the cautionary tales.

And my next thought?

So . . . who's going to play me in the next *God's Not Dead* sequel?

Falling for Science

I was born and raised in a very conservative church tradition. We followed the Bible to the letter. Our motto was "speak where the Bible speaks and be silent where the Bible is silent." We put a high premium on getting things right: there is one biblical way to worship, one biblical way to organize a church, and one biblical way to become a Christian. I essentially had a degree in proof-texting by the third grade.

I was taught from a young age that most churches are on the wrong side of these essentials. We wore our badge of biblical-ness proudly, standing firm against all the other churches who were in error.

It is no surprise, then, that my growing-up church and other churches with a similar theology are comfortable standing against the entire scientific community when it comes to evolution. We have the truth and the scientists are wrong, duped *en masse* by fossils and other frauds. Evolution is synonymous with atheism, end of discussion.

I fell in love with science in the seventh grade.

Hard to believe, but my junior high school did not require science for seventh graders; it was an elective. I planned to take seventh grade choir (it sounded fun), but my dad (who was a teacher in my school), thought otherwise. Unfortunately for me, the choir room was across from his classroom, and my very organized and disciplined dad did not care for the lack thereof he saw in the seventh-grade choir.

Seventh-grade life science was co-taught by a fun young man and a curmudgeonly (but funny) older man—they were quite a pair. We began with a tour of the animal kingdom, including dissections. Dissections were the real deal in the seventies, not the virtual, opt-out of blood-and-guts computer simulations available to the squeamish today.

In dissection labs, I discovered a whole new world of animals: sea stars, earthworms, crayfish, and really big grasshoppers. Now, I knew about these creatures, but I'm sure I never thought of them as animals. Animals are furry and cute and familiar, not exotic and headless like starfish. Oh, but I loved it. For the first time, I saw life on our planet as vast and varied and categorized in ways I never knew.

My high school biology teacher had a sarcastic sense of humor and he looked like the son-in-law on *All in the Family* (this passed for edgy in my high school). He liked to drop vaguely controversial comments like "No one ever said Adam and Eve looked like Mr. and Miss America." In retrospect, I'm pretty sure he accepted evolution. But this was small-city Texas in the seventies, and that might get you fired.

With a much deeper dive into animal classification, anatomy, and physiology, my awe and wonder returned. Some animals are very simple and primitive, with no tissues or organs, but are animals, nonetheless. Some animals have complex structures and systems very much like our own, and many others have versions, adaptions, and reworkings somewhere in-between. It is almost like (dare I think it?) these animals are connected in a much deeper sense, beyond the order of the syllabus. The primitive animals intrigued me most—dare I consider the unfolding, sequential nature of the fossil record?

I was on the high school debate team, and I lived (and still do) for the well-constructed argument. To this day, I retain the artifacts of that training: I ask too many questions in conversations and I love arguing both sides of an issue. Devil, here's your advocate, let's play. At some point in my teen years, I became aware of *apologetics*, the organized approach used primarily in religion to argue or explain a belief. The apologetic field in this case was "proving" the existence of God and the veracity of the Genesis account of creation. Apologetics is arguing and has the aura of science, and I loved it.

I took my love of biology to college, along with my questions and the apologetics. I graduated with a degree in biology from a Christian university, where we conveniently ignored the topic of evolution.

I knew lots of Bible facts, and I had an undergraduate's knowledge of biology. Armed with this impressive skill set, I tried to make the science fit Genesis. Let's see now:

in what order does Genesis say the animals appeared? Can we close one eye and squint and make the fossil record coordinate with the Genesis days of creation? Could a "day" actually be millions of years? Can we make those weird Bible animals like the leviathan into dinosaurs? Making science fit Genesis requires a lot of mental gymnastics, and I was working on a gold medal.

As a young adult, newly married and eventually with two little kids, the questions occupied a back burner, but my husband (a medical doctor) and I both loved Christian science apologetics. When our kids were just tinies, we dragged them to an Atlanta-wide event featuring a noted Christian apologist. He was a young earth creationist and the director of a big apologetics publishing house. I don't remember a lot of what he had to say (thanks to the tinies in tow), but he spoke eloquently about young earth creationism in a large public auditorium to a sparse crowd, including some vocal hecklers.

The Camel's Nose

About this time, I began reading apologist John Clayton, a geologist and physical science teacher. A former atheist, Clayton was a speaker and publisher of a monthly journal, *Does God Exist?* I promptly subscribed. His apologetic materials were unlike any I had ever read. Clayton directly addressed the geological evidence for the age of the earth, and he did not back down. Genesis 1:1 is an undated and untimed verse, according to Clayton, and as such, cannot be used to support a six-thousand-year-old earth.

I met John Clayton once at a speaking event. I do not agree with his conclusions regarding biological evolution, but I am forever grateful to him as a man of faith and a man of science who first told me geology is true, the earth is ancient, and Genesis can be read in a different way.

As with Darwin, evidence for an ancient earth was my gateway. The camel's nose was completely inside the tent and looking around for more. I read about evolution. I read about the evidence. I read about fossils. I read more about the age of the earth and the universe. The fuzzy, unspoken connections I made in junior high, high school, and college were finally given a vocabulary.

Even my Christian university alma mater was coming around. Unfortunately, the creationist I heard years before in Atlanta used his big platform and the purse strings of alumni to "take down" one of our beloved biology professors for the sin of mentioning evolution. Our professor lost his career, and my husband and I were heartbroken. My university persisted, however, and evolution is now taught as the foundational theory of biology.

The game-changer for me was Kenneth Miller's 1999 book, *Finding Darwin's God*. The book was a breath of fresh air and a challenge, all at once. Dr. Miller is an authentic, practicing Christian. Miller is also a noted cell biologist, researcher, author, and evolutionary biologist. In his book, Miller takes the claims of creationism, in all its forms (young, old, intelligent design), and without apology, overlays the science, the actual science. He doesn't flinch. And to tie it all together, Miller talks about what science can't tell us: What is right or wrong? What is good? What is evil? For Miller, these questions are answered by his Christian faith.

There was no turning back. Over the next two decades, there were many other men and women of faith and science who taught me through their books, their lectures, and their lives that I don't have to choose a side—faith or science. No more mental gymnastics to protect my interpretation of Genesis. Instead of eroding my faith, I found my faith growing in awe of a creator who was more unbound than I ever imagined.

A Question of Credibility

Few theological discussions among Christians arouse the level of heart-racing, blood-pressure-spiking defensiveness as a discussion of evolution. Who are you going to believe? The Bible, or a bunch of godless scientists?

What do you accept? What do you believe? I accept facts supported by empirical evidence, but the most important things in my life are the things I believe. I believe my family loves me, but I cannot prove it. Although evidence tells me I can reasonably and rationally believe in my family's love, I can't prove it using the scientific method.

Science does not answer all questions. The most important questions humans ask cannot be answered by science. I believe in a personal, loving God. I believe Jesus is who he said he was. I believe in the resurrection. I believe there is reasonable and rational evidence for these beliefs. These beliefs are the most important things in my life, but they are not science.

What happens when science-denial is a tenet of faith?

How can we expect people to believe us regarding things requiring faith (Jesus, the resurrection, miracles), when we deny observable, testable, and measurable science evidence? Are we credible?

I am a committed, practicing Christian who accepts the evidence for evolution and the age of the universe. I am a faithful member of an evangelical church. I take the Bible seriously. I believe every word of the Apostles' Creed and the Nicene Creed.

That's where I am now, but it's not where my story began.

I am a Christian.

I accept evolution and the descent of all life, including humans, from a common ancestor.

I accept the evidence for an ancient universe.

This book is by no means exhaustive. I've tried to avoid the "Gish gallop"[1] regarding both science evidence and anti-evolution arguments. I can't include it all, but I think it's a good start. My goal is to have a respectful conversation about the science of evolution and origins and the response of Christians to it.

2

Making Science Fit Genesis

It was an evolution of sorts in adhesive chrome-plated automobile accoutrement.

First there was the primitive Christian fish, simple and unadorned. The next permutation was the same little fish, but now sporting legs and feet and the name "Darwin" across its belly. The final model was the Christian fish, this time emblazoned with "Truth" in mid-swallow of the "Darwin" fish.

Ha! Top that.

The implication is unmistakable: religious faith and Darwin are incompatible, and more than that, Christianity completely trumps—no, *obliterates*—Darwin.

According to the Truth Eating Darwin model, here are your choices:

1. Reject a vast body of science evidence or
2. Reject God.

The Truth Eating Darwin model explains it all: serious Christians don't believe in evolution and serious scientists don't believe in God. In many Christian circles, particularly evangelical ones, science is suspect.

I have a box of trading cards, a dinosaur book, and a glossy *National Geographic*–style magazine, "regifted" to me by a friend. Published by the Institute for Creation Research, the materials are high quality and kid-friendly, with dinosaur facts and photos of giant insects and exotic animals. Woven within the photos and facts for science-loving kids is this clear message:

Scientists guess.

Scientists ignore facts.

Scientists are atheists.

Sometimes, science is downright feared. An ominous advertisement on a Dallas–Fort Worth "family-friendly" radio station warns: "Every time your child picks up a secular textbook they are exposed to agendas, propaganda, and inaccuracies."

Science isn't just wrong. Science is an enemy.

For the Christian who wants to love God with heart, soul, *and* mind, rejection of either science or God is an impossible choice. And for many Christians, overt rejection of science just feels wrong. After all, we are citizens of the twenty-first century. We love our technology and we depend on, and praise God for, advances in modern medical science.

For some people of faith, there is another path. Instead of an outright rejection of science, many find science in the Bible, and particularly in Genesis. Large and popular publishing houses, think tanks, and tourist venues with large staffs of researchers are dedicated to the goal of interpreting all science evidence regarding origins through the filter of a literal Genesis.

Making Science Fit Genesis

In most of prehistoric Europe, the sun dominated the cultural worldview. Structures similar to Stonehenge in Britain are found throughout Europe and appear to be built with an eye toward some sort of sun observation. Prehistoric Arctic Inuit, on the other hand, lived in a land of long dark winters. For the ancient Inuit, the moon was the central astronomical feature and the sun was a distant second place.

Astronomy is our most ancient science. It is no surprise humans have long studied the heavens, especially movements of the sun and features of the night sky. Practical needs drove this attention: charting growing seasons for crops, navigating on land and water, creating a calendar. Millennia before modern science, humans observed the heavens and wove their observations into their stories, their religions, and their cultures.

Ancient Greek astronomers were, in essence, philosophers. Influenced by Plato and Aristotle, Greek astronomers were mostly concerned with the nature of the heavenly bodies, the causes of their movements, and their influence on human beings. Alexander the Great's conquests into the Near East brought philosophical Greek astronomy into contact with the arithmetic-based astronomy of the Babylonians.

Classical Greek astronomy reached its apex in Ptolemy of Alexandria (AD 100–170). His *Almagest* was filled with geometry, charts, graphs, models, calculations, and tables. Ptolemy also included the longitude, latitude, and magnitude of over a thousand stars within their constellations.

Ptolemy's model of the universe placed the earth at the center of eight transparent spherical crystal shells, each nested concentrically inside each other like spherical Russian dolls. With a fixed earth at the center, the other shells were occupied by the moon, the sun, and the five known planets. Beyond these seven shells was an additional shell embedded with the stars. The eight crystal shells carrying everything in the known universe orbited in perfect harmony around the unmovable earth.

When the Roman empire fell, Greek thought (including Greek science) was lost to the western Christian world for centuries. Fortunately, classical Greek astronomical knowledge was absorbed into Islamic scientific culture.

In the early Middle Ages, wars and conquests between the west and the Islamic world exposed the Christian west to long-lost classical works like the *Almagest.* Early European universities, starved for real scientific insight, enthusiastically consumed the works of Ptolemy. Of course, the Ptolemy model of the universe is now obsolete, but in its day, it was the best science using the best instruments, observations, and mathematics. For the first time in fledgling European university centers, works of astronomy rose to the prestige and rank held by works of medicine and theology.

The academic delight, however, was short-lived. By the thirteenth century, European academia was in turmoil over Ptolemaic astronomy.

At issue was how to make Ptolemy's model of the universe "fit" into a literal reading of the Genesis creation account. Where is the "firmament"? Where are the "waters above the firmament"? Where is the distinction between the stars and the "heavens"? The solution was easy: Genesis is literal, so the model must be changed. Two additional spherical shells were invented in the Middle Ages to bring the model into line with the wording of the Genesis creation account.[1]

New instruments, better observations, and new data soon replaced the earth-centered Ptolemaic model with a sun-centered model. But this would not be the last time a forced fit between science and Genesis was attempted.

In the decade before Darwin published, French and British paleontologists and geologists were making remarkable geologic discoveries. Everyone knew rock formations of different kinds had their own unique assemblages of fossils. This observation was made independently and many times over by geologists all over Europe. In order to avoid any evolutionary ideas, many geologists and paleontologists (including the notable French scientist Baron Georges Cuvier) crammed multiple, separate creations into the one Genesis creation account.[2] Before long, twenty-seven separate creation events were stuffed into Genesis in order to preserve a literal reading and avoid any association with evidence for evolution or an ancient earth.[3]

In the twenty-first century, we are still trying to fit the square peg of science into the round hole of Genesis. Sometimes we try to make science fit Genesis. Sometimes we try to make Genesis fit science. Either way, the mental gymnastics required can be staggering.

Dinosaurs Are a Big Problem

Dinosaurs, being large (and awesome) are hard to ignore. Dino-crazy little kids are cheered on by doting parents, enthused by their offspring's early science inclinations. (Actually, a study found a correlation of high intelligence and dinosaur-obsessed kids).[4]

No doubt—dinosaurs are the stars of the paleontological world. Fossil skeletons of mammoths and giant sloths have their fans, but museum visitors crowd the dinosaur wings. Ken Ham, president of Answers in Genesis, recognizes the star power of dinosaurs: "We're putting evolutionists on notice: We're taking the dinosaurs back. . . . Evolutionists get very upset when we use dinosaurs. That's their star."[5]

It is no surprise that the question of where dinosaurs "fit" into Genesis is a common one.

Creationist children's books exposing the "truth" about dinosaurs include illustrations of men fighting a vicious T-Rex, hunters using a harnessed horned dinosaur to haul a captured tyrannosaur, and an idyllic scene of a young girl and her dinosaur pet. Stone figurines from ancient human cultures are offered as accurate models of dinosaurs.[6]

In the Genesis account, all land animals and humans were made on the sixth day of the creation week. Dinosaurs, being land animals, were therefore created on the same day as humans. Immediately we have a problem with geology.

In cases where geographic strata are intact, no humans or evidence of humans are found anywhere near dinosaur fossils. In fact, sixty-five million years pass between the extinction of the dinosaurs and the first modern human fossils. The timeline just doesn't work.

But Genesis is history. Genesis is literal. We have to make it work.

Lacking any human and dinosaur fossils in the same geologic layers, we turn to Biblical descriptions of "behemoths" and "leviathans" and "dragons".[7] Here's the logic: prior to the invention of the word *dinosaur*, these were the terms used by humans to describe the dinosaurs living among them.

And just like that, we make the science fit Genesis.

Noah's ark, however, poses the biggest dinosaurs-in-Genesis challenge. Noah was commanded to put two of each "kind" of land animal on the ark, which of course includes the dinosaurs. (Note: creationist literature defines "kind" as a generic group of similar animals.) Most commonly, creationists estimate fifty to sixty "kinds" of dinosaurs at the time of the flood.[8] We therefore need to load 100 to 120 dinosaurs on the ark, in addition to two and sometimes seven of each "kind" of every land animal on earth.

We must provide accommodations for all these land animals, plus food for a year-long voyage, plus fresh water for all, and of course, food and water for the people. And speaking of food, many dinosaurs as well as numerous other land animals are carnivores. The food supply must include herds of feeder animals and their supplies, also.

Housing and supplies for 100 to 120 enormous animals for a year pose an enormous problem. And *enormous* is not hyperbole—one group of dinosaurs, the titanosaurs, were the largest animals to ever walk the planet.[9] Even if some of the other representative dinosaur "kinds" were not large, many would be massive.

But Genesis is history. Genesis is literal. We have to make the science fit.

The solution? Baby dinosaurs on the ark. Occasionally it's not actual dinosaurs but dinosaur eggs; in some form, it's always baby dinosaurs on the ark.

The biologic and metabolic needs of enormous dinosaurs for a year? The physical space needed? A food supply for enormous animals, many of which are carnivores?

Baby dinosaurs solve all the problems. Baby dinosaurs make the science fit Genesis.

Evolution, God, and the Shrinking Third Path

The Gallup Poll has been reporting trends in American religious beliefs for decades, including beliefs regarding the origin of humans. From 1983 to 2019, one trend has held steady—the percentage of Americans who believe God created humans in their present form within the past ten thousand years. From a high of 47 percent in the 1990s, the percentage of Americans with strictly creationist beliefs is currently at 40 percent. Predictably, the majority of the respondents in the 40 percent were regular churchgoers.

One trend, however, was markedly different in the 2019 poll and is, in my opinion, quite concerning.

Of those who believe humans developed by evolution, Gallup differentiates between those who believe God guided evolution and those who believe God had no part. The percentage of Americans who do not believe God had any role in evolution has been steadily increasing and is currently at an all-time high (22 percent). Correspondingly, the percentage of Americans who see God in the process of evolution has been gradually *decreasing* since the late 1990s.[10]

As a Christian who accepts the evidence for evolution, the steady 40 percent holding a strictly young earth creationist view is disheartening. More troubling is the growing number of Americans who no longer believe they can accept science and also be a person of faith.

What is happening in our public conversations and in our shared collective impressions that implies a person who accepts evolution must reject religion, and conversely, a religious person must reject science?

Choosing Sides

April Maskiewicz Cordero is a biology professor at Point Loma Nazarene University. Her research agenda focuses on the challenges of teaching evolution to Christian students. The majority of Cordero's students have been told at some point prior to her introductory biology course that "science and faith are not compatible." At least 50 percent of Cordero's students report being told they had to choose between science and faith.[11]

Not surprisingly, many young adults get this message before they head off to college.

More troubling is the growing number of Americans who no longer believe they can accept science and also be a person of faith.

A comprehensive study of youth ministry in America found the most common questions about science asked of youth ministry staff and volunteers are about origins and evolution.[12] Unfortunately, the vast majority of youth ministers who attempt to discuss evolution with their kids must prepare their own materials. If published materials are used, the most commonly used resources are young earth creationist materials.

Many ministry resources carry a strong "how to send your kids off to college" theme. The most frequently cited young earth creationist resources make it clear: college professors *intend* to destroy your child's faith.[13] Biology, geology, and astronomy majors are specifically warned: professors are not to be trusted. Science majors are told "be careful who you tell" because some students have been "stopped from getting degrees" when they were outed as creationists. Helpful hints are given for writing papers and answering exams without showing your (creationist) hand.[14]

In popular culture, there are very loud, very prominent public voices on both ends of the faith/science spectrum. Popular creationist Ken Ham (and his many supporters) as well as evolutionary biologist and "new atheism" apologist Richard Dawkins (and his many supporters) are adamant: God and evolution cannot be friends.

Ham made his way into our public consciousness in his widely viewed and discussed 2014 debate with Bill Nye (the Science Guy). At the conclusion of the Ham and Nye debate, Ham was asked what science evidence would convince him regarding evolution. His answer? "Nothing." In defending the Christian perspective, Ham was unyielding—any science evidence supporting an ancient earth and evolution must be discarded in favor of a literal reading of Genesis.

Dawkins is an evolutionary biologist, researcher, and author. Dawkins is also an outspoken critic of religion (he considers all religion dangerous) and has been a guest on many non-science programs such as *Real Time with Bill Maher*, *The Daily Show* (with both Trevor Noah and Jon Stewart), and even had a guest spot on *The Simpsons*. Dawkins is so opposed to a reconciliation between faith and science he has publicly eviscerated National Institutes of Health director Francis Collins because Collins is an unapologetic Christian.[15]

Not to be excluded, television and film also weigh in. American's current favorite small-screen scientist is nine-year-old "young" Sheldon, first seen as adult Sheldon on the popular and long-running *Big Bang Theory*. Young Sheldon's good-natured pastor patiently entertains his science and religion questions, even when they come in the middle of his sermons. Here are our takeaways from *Young Sheldon*: churches believe Darwin was right regarding his belief in God, but wrong about evolution. And of course, there was no "big bang," according to the Bible and Sheldon's pastor. Pastor Jeff is adamant: there was no big bang, only God's word. (Of course, young Sheldon asks if that word was *kaboom*.)[16]

And if you don't believe young Sheldon, the popular *God's Not Dead* movie series drives the point home: atheistic professors stand on the side of science, faith-filled students stand with God.

The first movie in the *God's Not Dead* trilogy features an atheist professor challenging a Christian student to a public debate about the existence of God.[17]

The challenge begins when the professor requires all students to write the words "God is dead." The Christian student responds: "I can't do what you want. I'm a Christian." I can't help but think my two biology students filled out their notecards with this very scene in mind.

The debate focuses on the beginning of the universe and biological evolution. Of course, the professor argues for the big bang and evolution and the Christian student stands up for God.

The lead characters are straight out of Christian film central casting: the evolution-defending professor is obnoxious, rude, and joyful as he publicly humiliates the student. At one point, he glowers menacingly as the student enters his elevator. As the elevator door shuts, the professor hisses behind the student's head, "I made a mistake with you, letting you speak in front of my class and spew your propaganda." The Christian student is humble, earnest, well-spoken, and wins the respect of the other students.

When given his turn in front of the class, the student begins his defense: "For the last 150 years, Darwinists have been saying that God is unnecessary to explain man's existence and that evolution replaces God." The student continues discrediting evolution with a recitation of things Darwin "never addressed."

Of course, there's a happy ending. The student is voted "winner" of the debate and the professor eventually professes belief in God. But the film's win (and popularity) comes at a cost: misrepresenting the science and drawing an unnecessary line in the sand between science and faith.

Whether it's via church or popular media, the cultural message is clear: you have to choose—science or faith.

Apparently, Americans are listening.

Increasingly, they are choosing sides.

3

What Is Science?
The Nature of Science and EVO 101

Too many women were dying.

When Dr. Ignaz Semmelweis arrived for his new job in the maternity wards at Vienna General Hospital, he was immediately struck by the number of women who were dying of childbed fever. The year was 1847, and deaths during or following childbirth were unfortunately all too common. Still, Semmelweis was perplexed.

In the maternity ward where women were attended by male doctors and medical students, the death rate from childbed fever was a ghastly 29 percent. These women did not die giving birth or from birth complications, but instead were dying days after delivery.

In the maternity ward staffed by female midwives, the death rate from fever was only three percent.

Semmelweis was a doctor at the beginning of a golden age of medicine. Doctors were expected to have some scientific training and to approach patient care from a science perspective. Semmelweis, the physician/scientist, thought about possible explanations for the vast difference in fever death rates between the two wards.

Semmelweis first noticed birthing positions. Women who gave birth in the doctors' ward did so on their backs, while women in the midwives' ward gave birth on their sides. So, Semmelweis directed the women in the doctors' ward to give birth on their sides. The death disparity persisted.

Back to the drawing board for Semmelweis. Whenever a woman died in one of the wards, a priest walked the halls, mournfully ringing a bell. Semmelweis hypothesized that the post-mortem bell ringing so terrified the recovering women, they developed a fever, got sick, and died. Semmelweis ordered a stop to the bell ringing. Again, no change.

There remained one difference between the doctor-delivered ward and the midwife-delivered ward. Between deliveries, the doctors and the medical students dissected cadavers, sans gloves. When the time came to deliver a baby, the doctors and students might quickly rinse their hands in cold water . . . or they might not. After the baby was delivered, the doctors and students returned to their autopsies. The process was repeated throughout the duration of the doctors' shifts.

The midwives did not dissect cadavers. They just delivered babies.

Semmelweis's final hypothesis was this: bits of cadaver on the hands and under the nails of doctors cause childbed fever. As a test, doctors and medical students were ordered to wash their hands and instruments with soap and a chlorine solution after an autopsy and before delivering a baby. And what do you know? The death rate from childbed fever in the doctors' ward dropped to the same rate as in the midwives' ward.

Semmelweis tried to convince doctors across Europe to practice handwashing. Doctors were generally hesitant and were all too often offended by Semmelweis's charge that they, the doctors, were at fault. However, Semmelweis was not tactful and he publicly berated doctors who were hesitant to implement the practice.

Vienna General Hospital eventually gave up handwashing, and handwashing did not catch on in Europe. Semmelweis died at age 47 in an asylum, angry and mentally ill.

But then other doctors and scientists replicated Semmelweis's experiment and patient survival rates increased. No longer was handwashing just a hunch. Semmelweis's work was foundational to what would come to be known as the germ theory, and germ theory heralded the beginning of modern medicine.

Today, when a patient presents with an infection, doctors don't start from scratch. They don't consider foul odors or evil spirits as the cause of the illness. Doctors no longer consult the stars or planets or the patient's horoscope. No longer do doctors check for the balance of "humours" and bleed their patients if the blood is unbalanced. When doctors look for the cause of an infection or when researchers look for vaccines and treatments for new viruses, they start with germ theory. They start with the knowledge that infectious disease is caused by microscopic pathogens, and work from there.

Just a Theory?

When scientists use the term *theory*, they mean something quite different from a casual use of the term. For example, I have a theory that my Dallas Cowboys will win it all this season. That's a hunch, and probably just a hopeful guess.

A science theory is quite different. A science theory begins as an observation. Recall, Semmelweis observed disturbingly different death rates in the two maternity wards. He then proposed possible explanations for what he observed: birth position, bell ringing, hands with bits of cadaver clinging to them. A possible explanation for something observed is called a *hypothesis*. After testing each hypothesis, Semmelweis found only one supported by evidence: the dirty hands hypothesis.

After many years and many decades of study, after multiple scientists collect mountains of evidence and publish it for other scientists to critique, a hypothesis that continues to be supported time and time again reaches the rank of *science theory*.

When a hypothesis becomes a theory, are we done? Does it mean we know all there is to know? Of course not. As we learn new things and find new evidence, theories will be tweaked. But once an idea reaches the status of science theory, the foundations and fundamental principles are set. Will we discover new information about disease, adding to our understanding of germ theory? Of course, we will. But are we going discover that it actually *is* stars and planets and smelly air and *not* pathogens that cause disease? No.

That's the strength of a science theory. Science theories are the foundations of their field and the fundamentals aren't going to change. Can you image trying to do chemistry without atoms and molecules? Do you want to try physics without math? Gravitational theory, cell theory, atomic theory—all have been tested time and time again, all hold true, all are stable in principle.

Suppose you need surgery. Your surgeon comes into your hospital room and announces: "I'm in a hurry today, so I'm not planning to wash my hands or wear gloves for your operation. I'm also going to skip sterilizing the instruments. It's ok, after all, germ theory is only a theory."

How fast are you out of there?

Unfortunately, the casual meaning of the term *theory* is all too often applied to the theory of evolution. Usually this is in the context of "why can't we teach creation science or intelligent design in schools along with evolution? After all, evolution is 'only' a theory." Or, "evolution isn't a fact or law, it's 'just' a theory."

Actually, if we are to make a science term hierarchy, *theory* is at the top. Science theories rank *above* laws and facts because theories make sense of laws and facts. Theories knit laws and facts together into a coherent whole.

Like gravitational theory, germ theory, atomic theory, and cell theory, evolution theory is supported by decades of research. Evolution is foundational to understanding all of modern biology. Evolution principles guide what we do in environmental science and conservation of endangered species. Evolution principles guide our research in new diseases, new treatments, and antibiotic-resistant bacteria. Evolution principles guide our understanding of modern agricultural practices. Geneticist Theodosius Dobzhansky famously said, "nothing in biology makes sense except in light of evolution."

You may have questions about evolution, you may doubt it or reject it, but you cannot validly label it as "just a theory."

How Science Works

Lawyers are advocates. A good attorney defends the validity of her client's position. In a civil case, both opposing positions may be valid, but the attorney's job is to advocate for her client. Judges weigh both claims and render a subjective decision.

Science is different. Science wants to know what *can be known*. Science asks the question "what is true?" Science depends on reality—the natural world. Science works to examine evidence without a goal of "winning." Failed experiments are still successful—we know what the answer is *not*.

Science is self-correcting. After an unbiased examination of evidence, it is no shame to change your mind.

Prior to the 2015 fly-by of Pluto by the New Horizons spacecraft, scientists thought Pluto was an inert lump of ice—an inactive planet-wannabe tagging along with the cooler planets, trying to sit at the popular table. But all that changed with the New Horizons fly-by. We now know Pluto is geologically active, with high mountains and possible volcanoes. With new information, we changed our description of Pluto. People can still debate whether or not Pluto should be classified as a planet, but we can no longer say it is inactive.

Science is more than just a collection of facts. Science is a process. Science is what we use to understand the natural world.

Peer Review (or, Scientists Love to Prove Each Other Wrong)

Dr. Frances Arnold, winner of the 2018 Nobel Prize in Chemistry, began 2020 with a "bummer" of an announcement. On January 1, she tweeted:

"For my first work-related tweet of 2020, I am totally bummed to announce that we have retracted last year's paper on enzymatic synthesis of beta-lactams. The work has not been reproducible."[1]

The research mentioned was not her Nobel Prize–winning study, but the retraction was disappointing, nonetheless. Once published, other chemists found errors in her data. No other chemist could reproduce her findings. Even Nobel Prize winners are not immune to peer review. Others will show them to be wrong if they are, Nobel Prize or not.

Science is truth-seeking. Science does not advocate for a position for the sake of winning. Science seeks to know what is.

Science evidence is not considered valid unless it has been put before the broader body of scientists in the field of study. Studies must be published in peer-reviewed journals (reviewed by experts in the field) or presented at peer-reviewed academic conferences. But journals and conferences are just the beginning. Scientists have no qualms about disagreeing with each other or demonstrating errors in one another's research.

Science is truth-seeking. Science does not advocate for a position for the sake of winning. Science seeks to know what is.

A few years ago, I heard a lecture by Neil deGrasse Tyson. He took questions from the audience, and the inevitable question arose about the age of the universe and evolution. I'll never forget the message in his response: if anyone, anywhere, could demonstrate that what we know about the age of the earth or the evolution of life is incorrect, that individual would win a Nobel Prize. In other words, bring it on. Present your best evidence, let scientists in the field review it and see if your findings can be replicated by others.

Charles Darwin and Armadillos for Supper

The Royal Tyrrell museum in small-town Drumheller, Alberta, is a must-see if you love dinosaurs. The Tyrrell holds a wealth of enormous tyrannosaurs, elaborate horned and frilled dinosaurs (are these for real?), and herds of T-Rex-sized duck-billed hadrosaurs.

When I visited the Tyrrell, I first worked my way through the introductory rooms, dedicated to oil and gas exploration important to this part of Canada. In these rooms were nods here and there to some good chunks of petrified wood and other unintentional finds discovered in the search for fuels.

And there he was.

Not often do I literally stop in my tracks, but I did.

And whoa.

Right in the middle of the room was a dinosaur fossil like you've never seen before. Not skeletal remains. Not a two-dimensional imprint. Not even a dried up and mummified carcass.

But a 110-million-year-old dinosaur *almost exactly* as he appeared in life.

A *National Geographic* photographer called it the most impressive fossil he's seen in his life: "It was like a *Game of Thrones* dragon. It was so dimensional, like a prop from a movie."[2]

In life, this plant eater was eighteen feet long and weighed three thousand pounds. He was plated in spikes and also sported two impressive twenty-inch spikes on each shoulder. He was found by a mine excavator operator six years before his public debut in the Royal Tyrrell. This dino is a new species of nodosaur, an armored cousin to the club-wielding ankylosaurs. He is the 110-million-year-old new kid on the block.

Paleontologists think he was swept down river by a flood, then out to sea where he quickly sank, belly-up. He settled into the impact crater with his back supported and was covered in sediment. Minerals infiltrated his skin and armor. Mark Mitchell, the technician who spent thousands of hours prepping the specimen, called it one of the best preserved and most beautiful dinosaur specimens ever—"the Mona Lisa of dinosaurs."[3]

Standing just inches from this dinosaur, face to face, I was overwhelmed by these two thoughts: This animal is REALLY OLD and living things have changed A LOT.

Hello, Captain Obvious.

About 180 years ago, a young Charles Darwin had those same two thoughts on his famous round-the-world trip aboard the *Beagle*. In Argentina, Darwin ate a local delicacy: armadillo roasted in its shell. In this part of Argentina were also found fossils of a huge animal, a glyptodon.

Glyptodons are extinct. And glyptodons look just like enormous armadillos.

Darwin couldn't help but notice the little armadillos he was eating looked very much like the giant glyptodon fossils found in the same geographical area. Had one species replaced the other over immense amounts of time?

In Darwin's day, most people thought living things were static—what you see is what has always been. In other words, all living things were specially created by God in their present form, about six thousand years ago. Darwin, however, was not the first to suggest things might not have always been this way.

In the century before Darwin, geologists were beginning to understand that the earth was actually very, very old. In the eighteenth and early nineteenth centuries, fossil discoveries indicated that some species had died out. Other naturalists in Darwin's day thought life on earth may have changed, including Alfred Russel Wallace, who almost scooped Darwin on the whole evolution thing.

If Darwin was not the only or even the first to suggest that life had evolved, why is his name synonymous with evolution? Darwin was the first to advance the conversation in three important areas:

- Darwin proposed a mechanism by which evolution occurred: natural selection.
- Darwin incorporated the idea of deep time: evolution of species occurs over millions of years.
- Darwin proposed a tree of life model: all living things are related to each other.

In other words, Darwin was the first to advance an idea of *how* change happened.

Darwin's Big Idea

Darwin recognized that the struggle to survive is ever-present in nature. The creatures that survive are the ones having traits most suited for their environment: some are skillful killing machines, making the most of available prey. Others are skilled in hiding or camouflage and thus avoid *being* the prey. Other creatures can live in extremes of hot or cold, while others have beaks or teeth or mouths suited for the available food sources. Although the struggle could be savage, Darwin saw the process as creative, the result of which was "endless forms, most beautiful."[4]

Darwin understood the starting point for change was variation in traits. Even the tiniest variation in a trait can make the difference between survival and death. Over many generations, tiny variations accumulate, and a new species branches off in its own direction. This is Darwin's big idea: evolution by natural selection.

We know now what Darwin never knew—variations in organisms originate at the molecular level, in the genes. Changes in genes, or mutations, increase the amount of variation in a population. Genetic mutations can be thought of as "copying errors"—they are random and happen all the time. Often, these copying errors in DNA are caught and corrected by molecular copy-editing. Some mutations escape correction and can be harmful. Most mutations are neutral. Occasionally, a mutation is beneficial to an organism.

Genetic mutations occur in an individual by chance, but evolution does not happen in an individual. If a mutation provides some sort of an advantage, the mutated gene will increase in a population. If a mutation is harmful, it will disappear from the population. Natural selection really boils down to this: does this trait make an individual more fit for the environment? If so, the individual is more likely to survive, find a mate, and leave offspring in the next generation. If a trait makes an individual less adapted for the environment, the individual is less likely to survive, find a mate, and leave offspring.

Natural selection does not mean perfection. The term "survival of the fittest" is a misleading and inaccurate definition of natural selection. "Survival of the fittest" implies that an individual is the strongest or fastest or most talented. An individual does not have to be the best in order to be "fit." Natural selection is *not* survival of the fittest. Natural selection is survival of the "fit enough": fit enough to survive, find a mate, and leave offspring.

Evolution is often derided as a "random" or "chance" process. Skeptics shake their heads and ask how blind-chance events could produce the beauty and complexity we see in nature. On the contrary, evolution by natural selection is anything but random. Gene mutations are random, but whether or not a mutation takes off in a population is definitely not random. Species evolve because their traits make them fit for the environment. Traits that make a species less fit lead to its extinction. Natural selection is not a by-chance, roll-of-the-dice proposition.

It is amazing how much Darwin understood about natural selection, given he was born in 1809. Charles Darwin lived, worked, wrote, and died before anyone had ever heard of genes or chromosomes or DNA, much less modern molecular genetics. And because Gregor Mendel's work on the heritability of traits was not recognized until the twentieth century, Darwin never heard of him, even though they were contemporaries.

Yet, it is common for evolution critics to proof-text the failure of evolution using the writings of Charles Darwin. Countless apologists throw modern questions Darwin's way, and in each case, Darwin fails to answer or even address the questions.

Darwin died in 1882, seventy-one years before the structure of DNA was determined. There are countless questions left unanswered by Darwin. He simply did not have the tools to know. Darwin did not have the tools to answer, but twenty-first-century science does a really good job of it. No one suggests we throw out gravitation theory because seventeenth-century Isaac Newton never addressed the questions of twenty-first-century physicists. Despite a nineteenth-century framework, the bedrocks of Darwin's theory—evolution by natural selection over millions of years and the common ancestry of all living things—remain unchallenged.

Evolution's Toolkit

The Irish potato famine decimated the population of Ireland in the nineteenth century. The circumstances leading to this tragedy were multifaceted, but a crucial component was the potato itself. The only potato planted in Ireland at the time was the "lumper," a variety with no genetic variation. Every lumper potato was a clone of all other lumpers. When disease struck, the potato crop was destroyed. Had there been genetic variety in the potato crop, it is likely some potatoes would have carried genes for disease resistance. But lumpers were clones, so the potato crop was wiped out.

The lumper story illustrates an essential element of evolution: natural selection can only act on what is already there. Potato plants cannot create a new resistance gene from scratch. Potato plants cannot "try" to evolve a new gene in response to the disease. Natural selection can only select from genes already present in the organism. If a resistance gene is present in the potato population, plants with the gene will survive and leave offspring. All plants without the resistance gene will die and leave no offspring.

Now to 2014 for a similar story with a much different ending.

A research lab was studying *P. fluorescens*, a bacteria species commonly found in soil. *P. fluorescens* fixes nitrogen, making nitrogen in the soil available to plants.

Specifically, the researchers wanted to know what

would happen if the bacteria lost the ability to move about in their environment. The researchers knocked out the gene for growing a flagellum, a little outboard motor-like structure used for swimming.

The resulting bacteria colonies could not swim. Bacteria that cannot swim cannot get food once they have consumed all the nutrients directly adjacent to them. Therefore, flagella-less bacteria must be fed. The feeding job fell to the graduate assistant, who unfortunately, forgot about it for several days.

Fearing the worst, the researchers examined the Petri dishes. As expected, most of the bacteria were dead. Quite surprising, however, were the few small colonies that lived.

An even bigger surprise awaited. The surviving colonies had regrown flagella! Not as big or as strong as the original flagella, but good-enough flagella to swim in a Petri dish and find food. Lacking a flagella-building gene, how did the surviving bacteria re-grow their flagella?

The researchers compared the survivors and the non-survivors and found the answer in another gene.[5] This gene, however, had nothing to do with building flagella. Instead, it was a gene for building a nitrogen-regulating protein.

Turns out, the nitrogen-regulating protein is somewhat similar to the flagella-building protein, by about 30 percent. Here is the key—the survivors had a *mutated* version of the nitrogen-regulating gene. This mutation caused an *over*production of the protein. An overproduction of a protein similar to the flagella-making protein was just enough to grow smaller, but still functional flagella.

But the story gets better. In the natural soil environment, the mutation greatly reduces nitrogen regulation—definitely a disadvantage for nitrogen-regulating bacteria. In the natural environment, bacteria carrying the mutation are understandably rare.

However, we are growing bacteria in a Petri dish, not soil. The environment has changed. Bacteria without the mutation die. Bacteria with the mutation regrow (smaller) flagella and survive.

What can we learn about evolution from the sagas of the lumper potatoes and the flagella-less bacteria?

Environmental pressures drive evolution. When there is a change in the environment, populations must adapt or die. When a potato disease struck, there were no resistance genes in the population. The potato population was destroyed.

Without flagella, bacteria must adapt or die. A few bacteria had a version of a gene that was repurposed to grow smaller but functional flagella. Most of the bacteria died, but a subpopulation with a new kind of flagella survived in the Petri dish environment. All descendants of the subpopulation will have the new kind of flagella.

Natural selection can only act on the genetic toolkit already present in a population. Genes are repurposed. Genes are turned on or off. New combinations of genes result in dramatic changes in a species. Nature is thrifty and evolution is a tinkerer. Species must adapt or die. Today's mutation may be the key to survival in thousands or millions of years.

Two Sisters and an Elvis Convention

People resemble one another for different reasons. Consider two biological sisters: both have dark brown hair, green eyes, and a slight frame. The sisters aren't identical, but the resemblance is unmistakable. The sisters resemble each other because they are related. The sisters resemble each other because they share recent common ancestors.

On the other hand, people might resemble each other for a different reason. Consider two attendees at an Elvis impersonators convention: both have long sideburns and wavy black hair, and both are sporting a rhinestone-studded jumpsuit.

The similarities between the sisters are due to common ancestry, but the similarities between the two Elvises (Elvi?) are due to, well, other reasons.

The environment is a powerful driver of evolutionary change in living things. A change in the environment led to extinction of the lumper potatoes. A change in the environment drove the evolution of a new kind of flagella in the laboratory bacteria.

Environmental pressure is such a powerful driver of evolution, we find very distantly related animals looking very much alike. Consider these three very different animals: a shark, a dolphin, and an ichthyosaur (an extinct reptile). All three live (or lived) in the ocean, but one is a fish, one is a mammal, and one is a reptile.

All three are subject to the environmental pressures of a marine life. Do you need to move aerodynamically through the water? You have a streamlined body shape. Do you need to swim fast and quickly change directions in water? You have fins and flippers.

Evolution solved a similar problem (living in an ocean environment) in a similar way (streamlined bodies, fins and flippers) in three different animals. The body similarities of a shark, a dolphin, and an ichthyosaur were not inherited from a common ancestor. The body similarities of a shark, a dolphin, and an ichthyosaur evolved due to common environmental pressures. Of course, all three animals share a common vertebrate ancestor way, way back, but their common ancestor did not have a streamlined body, fins, and flippers.

We see evolution solving similar environmental problems in similar ways all over the planet. Australia is home to a unique population of mammals. All Australian land mammals are pouched animals, marsupials like kangaroos and koalas. The rest of the planet is populated with non-pouched placental mammals.

Although Australia (being a huge island continent) is physically separated from the rest of the planet, Australia has many of the same environmental conditions found all over the world. We find an astounding collection of Australian marsupials looking and living very much like placental mammals in places far from Australia. Marsupial sugar gliders look and live like placental flying squirrels. Marsupial Tasmanian devils look and live like placental wolverines. Marsupial kangaroos look and live like placental Patagonian cavies. In each case, evolution adapted very different animals to common environmental pressures in the same way. Again, marsupials and placentals shared a common mammalian ancestor long ago, but their common ancestor did not have the traits we see in modern marsupials and placentals.

Species must adapt to problems presented by the environment or die.

4

Where Are You Camping?
A Look at Beliefs

Rey is my favorite Star Wars character. Confession: I did not act my age and dressed as Rey when I saw Episode 8 in the theater. Imagine my disappointment when *BuzzFeed*'s "Which Star Wars Character Are You" quiz determined that I am the Tauntaun Luke sleeps inside ("you smell terrible, but people would be lost without you").

We like personality inventories. Whether it's the Meyers-Briggs, the Enneagram, or a BuzzFeed quiz, it can be informative (or even instructional) to see how one's personal traits, preferences, and beliefs fall into recognizable patterns. It is a rare person, however, who fits conclusively into a single inventory grouping without spilling over a bit into multiple other groupings.

Likewise, most people will not fall completely within the parameters outlined by any of the think-tank groups for "young earth creationism," "old earth creationism," "intelligent design," "theistic evolution/evolutionary creationism," or even "atheistic naturalism/scientism." Boundaries are fluid because the principal writers and influencers in each camp often vary slightly from others in the same camp.

My attempt to define terms acknowledges the nuances in views and positions, even within closely aligned camps. What follows are summary principles underlying the major think tanks, publishers, and influential voices regarding science and religion in the creationism camps (both young and old earth), intelligent design, theistic evolution/evolutionary creation, and naturalism.

One more caveat: in order to be precise, I use the terms *creationist* and *creationism* to identify certain associations and perspectives. There are, without a doubt, individuals who accept the evidence for evolution *and* believe God is the author and sustainer of all that is and regard God as the creator. As I define my terms, I use *creationist* and *creationism* to identify a specific viewpoint regarding the origins of the physical world.

Young Earth Creationism

Major Voices

The two most prolific publishers of books, articles, web-based resources, and church/homeschool curriculums from a young earth creationism perspective are Ken Ham's Answers in Genesis[1] and the Dallas-based Institute for Creation Research.[2] Both groups have impressive visitor destinations associated with their organizations: the Ark Encounter in Williamstown, Kentucky, and the Creation Museum in Petersburg, Kentucky, are attractions associated with Answers in Genesis. In 2019, the Institute for Creation Research opened a striking new museum on their Dallas campus, the Discovery Center for Science and Earth History.

Eric Hovind of *Creation Today* is a new face in young earth creationism.[3] He was recently featured in *We Believe in Dinosaurs*, the 2019 documentary about Ham's Ark Encounter attraction.[4]

Apologetics Press (Montgomery, Alabama), begun as a publishing house for young earth creation materials, is also a source for publications and curriculums. In addition to publications, Apologetics Press offers creationist summer camps and event speakers.[5]

Science Evidence

Georgia Purdom, a molecular geneticist with Answers in Genesis, succinctly articulates the young earth perspective regarding science evidence: "What I believe isn't based on the evidence, it's based on the Bible."[6]

Young earth creationists (YEC) have a singular approach regarding science facts: facts must be interpreted, and the Bible is the only lens through which all facts should be filtered. According to young earth creationists, the Bible provides a literal, firsthand, and historically accurate account of the beginning of the universe and life. Therefore, if there is a conflict between a literal Bible reading and science evidence, the Bible is always correct, and the science is always wrong. For example, the fossil record conflicts with the order of creation given in Genesis: fruit trees, followed by sea creatures and birds, then finally land animals. Because the order given in Genesis is correct, modern interpretations of the fossil record are, by default, incorrect.

While it is not an exaggeration to say to that the overwhelming majority of biologists accept evolution, there are working biologists (in the extreme minority) who reject it.[7] When biologists reject evolution, however, it is for religious reasons, not a lack of science evidence.

Todd Wood exemplifies this conundrum. Wood holds a PhD in biochemistry and is the founding president of the Core Academy of Science, a source for creationist resources. Wood ignited controversy on Core Academy's website with his blog titled "The Truth About Evolution":

> Evolution is not a theory in crisis. It is not teetering on the verge of collapse. . . . There is evidence for evolution, gobs and gobs of it. . . . There is no conspiracy to hide the truth about the failure of evolution. . . . Creationist students, listen to me very carefully: There is evidence for evolution, and evolution is an extremely successful *scientific* theory. That doesn't make it ultimately true, and it doesn't mean that there could not possibly be viable alternatives. It is my own faith choice to reject evolution.[8]

Young earth creationists agree: science facts, when interpreted correctly, will always confirm the Biblical account.[9]

Age of Earth and Universe

Young earth creationists believe the creation account in Genesis describes six literal, 24-hour days. The entirety of the universe and all life appeared within the six-day timeframe.

Young earth creationists believe the universe is between six thousand and ten thousand years old, with definite leanings toward the six-thousand-year mark. This timeline is calculated using the genealogies in Genesis, chapters 5 and 11. Based on the genealogies and assumed lifespans of Adam and the others listed in Genesis, creation is dated at about six thousand years ago and Noah's flood about four thousand years ago.

Young earth creationists find modern methods of dating tree rings, ice rings, and rocks to be faulty because they do not confirm a six-thousand-year-old earth. Likewise, astronomical evidence (for example, light traveling from distant stars) is interpreted assuming a six-thousand-year-old universe.

Creation Week

All life was specially created. No natural processes were involved. All organisms, including humans, were created instantly, separately, and in their present form, fully mature and functional.

Young earth creationists concede a point, however, regarding variations in basic "kinds." Young earth creationists believe organisms were created with the potential to give rise to multiple varieties within a generic "kind"—for example, lions, tigers, and domesticated cats are derived from a specially created "cat kind." Therefore, Noah only needed to take representative "kinds" aboard the ark. (Note: the concept of "kind" is unique to creationism and is not a concept found in modern biology.)

Young earth creationists believe all animals, including fully formed and developed carnivores, were created as vegetarians. No animal ate meat until after the sin of Adam and Eve.

Humans

The first man, Adam, was specially created, fully formed and mature, from dust. The first woman, Eve, was specially created, fully formed and mature, from Adam's rib. Humans share no ancestry with any other animal but were fully human in form and nature from their creation.

All humans who have ever lived are directly descended from this original pair. The eight survivors on the ark were direct descendants of Adam and Eve, and all humans since the flood descend directly from those eight.

Noah's Flood

Uniformly, young earth creationists believe geological evidence found in the earth's crust (including the fossil record) is primarily due to a catastrophic global flood, as described in the Noah account. Young earth creationists believe floodwaters covered the entire surface of the earth, including the highest peaks. Noah's flood is responsible for many of the geological features of earth, including the Grand Canyon. Furthermore, young earth creationists reject any suggestion of a localized flood.

According to young earth creationism, all of the earth's land animals, birds, and many sea creatures (those not saved on the ark) died in the flood. All humans on earth, except for the eight aboard the ark, also died in the flood.

Theology

Young earth creationists read Genesis as a literal, firsthand account of historical events, accurate in timeline, sequence, and details. Genesis is not poetry, allegory, myth, or any genre other than history. Young earth creationists believe Bible inerrancy demands a literal, historic Genesis.

Without a literal Genesis (including a belief in a young earth), the Bible cannot be trusted in any other aspect of faith. A plaque at the Ark Encounter reads: "If I can convince you that the flood was not real, then I can convince you that heaven and hell are not real."[10]

From a young earth creationism perspective, a literal Genesis is essential to Christian theology. The theology of sin, death, and the saving gospel of Jesus is dependent on a literal reading of Genesis.

Old Earth Creationism

Major Voices

Reasons to Believe (Covina, California) is by far the most recognized resource center for old earth creationism (OEC) events, speakers, authors, media, and publications.[11] Reasons to Believe founder, Hugh Ross, is the author of *Creation as Science: A Testable Model Approach to End the Creation/Evolution Wars* and is the face of old earth creationism.[12]

John Clayton's Does God Exist? (South Bend, Indiana) is a smaller old earth creationist operation.[13] Does God Exist? publishes a website and a monthly newsletter. As an earth science teacher, Clayton has long advocated for an old earth within his creationist perspective.

Science Evidence

Old earth creationists approach science evidence from a "two revelations" model: God's revelation in nature and God's revelation in the Bible. Regardless, any conflict between science and the Bible is awarded to the Bible. Fazale Rana of Reasons to Believe puts it this way: "It's not to say that there isn't evidence for common descent; there is. Still, even with this evidence, I prefer old-earth creationism."[14]

Within this model, old earth creationists believe science evidence will ultimately agree with the Bible.[15] For example, old earth creationists believe the meaning of the Genesis word *create* is consistent with the scientific description of the big bang.

Old earth creationists reject evidence for common ancestry of all life. Old earth creationists believe that God miraculously intervened billions of times throughout earth's history to separately create each and every species.

Age of Earth and Universe

As the descriptor implies, old earth creationists accept science evidence for the age of the universe (about 14 billion years) and the earth (about 4.6 billion years). Accepting the science evidence for the age of the earth and universe is a source of continuous conflict between the young and old earth creationist camps.

Old earth creationists usually reconcile the geologic evidence with a seven-day creation week using the "day-age" approach advocated by Hugh Ross. According to Ross, the Genesis word for "day" is best interpreted as an extremely long, but finite time period. In other words, the creation week was not composed of six literal 24-hour days, but billions of years divided into six periods of creative action.

Other old earth creationists separate creation into a time "before" Eden and "after" Eden, allowing for a gap of billions of years between the "beginning" and the Genesis creation week.[16] While the creation week is believed to be a literal seven-day event, there are billions of years separating Genesis 1:1 and Genesis 1:2.

Creation Week

Old earth creationists believe the days of creation as described in Genesis are a chronological account of the appearance of life on earth. In a billions-of-years-long creation "week," God specially and separately created each of the millions of species, both extinct and living. Old earth creationists reject all aspects of biological evolution. Any evidence indicating common ancestry between species is the result of a single Creator utilizing common designs and patterns.

Humans

Old earth creationists accept the evidence of the fossil record, including premodern humans like Neanderthals, Denisovans, and the Australopithecines. However, premodern humans are considered animals and not spiritual beings.

Adam and Eve were a literal, historical couple. They were separately and specially created and share no ancestry with any other creature. Old earth creationists believe all humanity descended from Adam and Eve.

Noah's Flood

Old earth creationists believe Noah's flood destroyed all humans and animals not saved on the ark. In a major departure from young earth creationists, many old earth creationists do not believe the flood was global in extent. Instead, the flood was limited to areas of human settlement.

Theology

Old earth creationists believe Genesis is without error historically or scientifically. Old earth creationists consider their interpretation of Genesis to be a literal interpretation.

Old earth creationists find support in the margin notes in the popular *Scofield Reference Bible*, published in 1917. In a center margin note for Genesis 1, the date for creation is given as 4004, BC. Scofield clarified the date in the study notes as the point at which the "Eden" story occurred. Historically, old earth creationists often claim scriptural authority for a period of immense time prior to the creation week.[17]

Like young earth creationists, old earth creationists consider the fall of a literal Adam to be an essential Christian doctrine. Unlike young earth creationists, old earth creationists believe Adam's fall resulted in death for humans *only*. In other words, non-human life (plants, animals) regularly suffered death before the fall.

Intelligent Design

Intelligent design (ID) is not really a view separate from creationist views, rather, it is a way of defending such views. Although intelligent design proponents steadfastly resist the "creationist" label, creationists regularly appeal to design and the failure of Darwinian evolution, both hallmarks of the intelligent design model.

Major Voices

The Discovery Institute (Seattle, Washington) is without a doubt the hub of intelligent design resources.[18] The Discovery Institute is a think-tank and policy center featuring the leading advocates of the intelligent design model. Within the Discovery Institute is housed the Center for Science & Culture. The Center for Science & Culture produces curriculums for religious schools, homeschools, and churches, as well as web-based media and documentaries.

The most prolific writers of intelligent design books and articles are Michael Behe, Stephen C. Meyer, and William A. Dembski.

Sean McDowell, son of legendary Christian Apologist Josh McDowell (*Evidence That Demands a Verdict*) is the face of rising-generation intelligent design advocates. Along with his father, McDowell released an updated *Evidence* (2017).[19] He also cowrote a book about intelligent design with William Dembski of the Discovery Institute. McDowell hosts the *Think Biblically* podcast and is a frequent speaker for apologetics productions.[20]

Science Evidence

The intelligent design model uses philosophical arguments in the context of scientific terms. Because the focus of intelligent design is on the complexity of biological systems and structures, intelligent design literature can be quite technical.

The centerpiece of intelligent design research is the concept of "irreducible complexity": unguided (natural) processes could never produce the complexity and machine-like qualities we see in biological systems and structures. Remove any part of the system or structure, and the system or structure is useless. Therefore, complex systems are specially and intentionally designed and are irrefutable evidence of a designer.

Intelligent design also focuses on the "unexplained" in science. Where there is a gap in science knowledge, intelligent design advocates ask, "is it actually explainable?" Intelligent design emphasizes purported "weaknesses" in evolution theory.

Intelligent design advocates consider ID to be a science theory:

> Intelligent design (ID) is a scientific theory that employs the methods commonly used by other historical sciences to conclude that certain features of the universe and of living things are best explained by an intelligent cause, not an undirected process such as natural selection.[21]

The absence of design research in peer-reviewed science journals is considered a product of bias against intelligent design by mainstream scientists.

Age of Earth and Universe

The intelligent design model accepts the science evidence for a billions-of-years age for the earth and universe.

Creation Week

Because intelligent design is described by advocates as a science theory, the intelligent design model does not make any claims regarding the creation week as recorded in Genesis.

The intelligent design model accepts the fossil record as an accurate record of the appearance of life on earth. According to the model, most groups in the fossil record appear abruptly and fully formed. Intelligent design allows for the possibility of common ancestry, but assumes the descendants are designed and not the result of evolution.

Intelligent design accepts some aspects of "change over time" yet rejects unguided natural selection as the source of variation we see in life.

Humans

Intelligent design makes no claims regarding a literal or historical Adam and Eve.

Intelligent design advocates regard the fossil history of early humans with skepticism. Early hominins like "Lucy" and some early members of the *Homo* genus are considered apes, not human ancestors. It is possible, according to intelligent design, that later members of *Homo* (like the Neanderthals) were human.

Intelligent design advocates believe the modern human genus, *Homo*, appears abruptly in the fossil record.

Noah's Flood

Intelligent design does not make any claims regarding Genesis, so Noah's flood or "flood geology" are not addressed by the model.

Theology

Intelligent design is described by its advocates as a concept completely separate from any religious tradition and claims no allegiance to any sacred book, including the Bible.

Although God is not a formal component of the intelligent design model, intelligent design is explicit: undirected natural causes (like evolution) cannot explain what we see in the universe and in life. The ID model attributes the natural world to an unnamed "intelligent designer." Without a doubt, the God of the Bible is assumed to be the designer by the overwhelming majority of intelligent design proponents.

Theistic Evolution/Evolutionary Creationism

Major Voices

Many people of faith who also accept the evidence for evolution describe themselves simply in these terms, without necessarily identifying with a particular group.

However, the terms *theistic evolution* and *evolutionary creationism* are both found as descriptions for this view.

The largest advocacy group for theistic evolution/evolutionary creationism is BioLogos (Grand Rapids, Michigan).[22] The BioLogos website curates, distributes, and reviews articles, books, and events supporting theistic evolution/evolutionary creationism. BioLogos was established in 2009 by Dr. Francis Collins, current director of the National Institutes of Health and former director of the Human Genome Project.

Collins is at the forefront of thought in theistic evolution/evolutionary creationism. Winner of the 2020 Templeton Prize, Collins has long advocated for the integration of science and reason.[23] In his bestseller, *The Language of God,* Collins defends his acceptance of both evolution and Christianity.[24]

In addition to Collins, cell biologist and author Kenneth Miller (*Finding Darwin's God* and *Only a Theory*)[25] and geneticist Dennis Venema and Bible scholar Scot McKnight (*Adam and the Genome*)[26] are authors of bestselling books defending theistic evolution/evolutionary creationism.

Science Evidence

Theistic evolution/evolutionary creationism accepts all science evidence regarding the universe and life. Evolution by natural selection and the descent of all life from common ancestry are both accepted. Proponents of theistic evolution/evolutionary creationism view science evidence as a valid revelation from God, a way to understand the *how* and *when* of creation.[27] Science evidence provides insight into how God creates, orders, and sustains the natural world.

Age of Earth and Universe

Theistic evolution/evolutionary creationism accepts the science evidence for an ancient age for the earth and universe.

Creation Week

Theistic evolution/evolutionary creationists affirm God as the creator and original cause of the natural world.

The Genesis creation account is written in the genre of ancient Near Eastern creation stories, of which there are several. Using a format common to cultures surrounding Israel, Israel tells its version, with God as the cause and at the center of the story.

Theistic evolution/evolutionary creationists read Genesis as theology, not as science or a literal historical account.

Humans

Theistic evolution/evolutionary creationism accepts the science evidence for common ancestry of humans with all life. Theologically, theistic evolution/evolutionary creationists believe God endowed humans with his image and has a unique relationship with them.

Theistic evolution/evolutionary creationists hold varying opinions regarding whether or not Adam and Eve were factual, historical people. However, Adam and Eve are not accepted as the genetic ancestors of all humans.

Noah's Flood

Theistic evolution/evolutionary creationists reject all arguments of a universal flood, "flood geology," or the descent of all life (human and others) from a few on the ark.

Many theistic evolution/evolutionary creationists read the Noah story as part of the genre of ancient Near/Middle Eastern flood stories, of which there are several. Like the creation story, the Noah story is read in the context of like-culture stories, but with God as the central figure in the story. Opinions vary as to whether Noah's flood was actually a localized historical event or a story with a theological message, or a combination of both.

Theology

Theistic evolution/evolutionary creationists are not deists. Most believe in an active and involved God who works miracles, including the resurrection of Jesus Christ.

There is a wide range of beliefs among theistic evolution/evolutionary creationists regarding the nature of the Bible. Theistic evolution/evolutionary creationists are found in protestant, Catholic, Orthodox, and a growing number of evangelical traditions, as well as in non-Christian traditions.

Overall, the Bible is viewed as authoritative for faith and life. Views differ regarding the concept of inspiration, but in general, inspiration is not seen as inconsistent with an "ancient genre" approach to Genesis.

Theistic evolution/evolutionary creationists generally believe science is limited to explanations of the natural world:

> When someone claims that the Bible answers a scientific question, and another claims that science answers a question about God, the conflict immediately flares up. Many conflicts become enflamed because participants forget that Christianity and science do generally address very different questions.[28]

Naturalism and Scientism

Major Voices

Naturalism is a term used to describe the philosophy or world view that a natural, physical existence is all there is—nothing exists beyond nature, nothing is supernatural.[29] Scientism is a closely related concept, a view that scientific inquiry trumps all other forms of inquiry.

Jerry Coyne is an evolutionary biologist and author of the bestseller, *Why Evolution is True* (2009). Coyne's 2016 bestseller, *Faith vs. Fact: Why Science and Religion are Incompatible* introduced Coyne to readers beyond the scientific community. Coyne pulls no punches: "science and religion are not only in conflict—even at "war"—but also represent incompatible ways of viewing the world."[30]

There is no fiercer contemporary voice for the naturalism/scientism worldview than eminent evolutionary biologist Richard Dawkins (*The God Delusion*, 2006).[31] Dawkins considers scientists who profess faith to be hopelessly inconsistent and untrue to the scientific method: "Once you buy into the position of faith, then suddenly you find yourself losing all of your natural skepticism and your scientific—really scientific—credibility."[32]

Both Coyne and Dawkins are relentless: religion and science are dipole opposites. Interestingly, most young earth creationists agree with Dawkins and Coyne—acceptance of evolution is considered irreconcilable with Christian faith and belief in the Bible. In this regard, Dawkins, Coyne, and Ken Ham are unlikely allies.

Leaving Camp Can Be Complicated

Fury broke out in East Kilbride, Scotland in the fall of 2013 when a local evangelical church gave elementary school children religious books as part of a chaplaincy program. Pilfering a Facebook photo of one of the church's young ministers (an American missionary) painted up as the pirate Jack Sparrow, a local tabloid published the story of the outraged community.[33] The story about the church was quite over the top (as tabloids tend to be) and painted the group as a scary American cult in which the minister dresses up like Johnny Depp and leads little kids astray.

The offending books? *Exposing the Myth of Evolution* and *How Do You Know God Is Real?* In his defense of the books, the minister of the East Kilbride church said, "We believe the teachings of the Bible, which tell us evolution is a myth."

For those of us steeped in inerrancy traditions, moving away from a creationist camp is complicated. When the Bible is literally and historically true and any apparent contradictions can be explained, evolution is, by default, false. End of story.

Still, even the most inerrantist among us allows the Bible to be, in some places, truth without being literally true. Case in point—New Testament parables. Coming out of a creationist camp doesn't require jettisoning the Bible, but it does require looking at Genesis in a different way.

New Testament scholar N. T. Wright finds the points of conflict and dissonance in the Bible drive him to a much deeper understanding of Scripture:

> One of the funny things about the Christian faith, and indeed about the Bible, is that it seems to be, as it were, designed that every generation has to chew it through afresh. We can none of us live on what was done before, because the culture is always changing, and that has always been so. The language is always changing. The pressure points for people are always changing. And again and again. And this is not just in our generation, every generation has found this, the way that people said things before seem to go stale on you.[34]

"Chewing through" Scripture afresh is not a novel, twenty-first-century concept. In light of the new thing God had done in Jesus the Messiah, all four gospel writers as well as Paul refocused Israel's Scriptures around Jesus.[35]

What to do, then, when science and a literal reading of the Bible conflict? For most Christians, tossing out the Bible is unacceptable.

Every one of the Bible writers lived long before modern science existed.

Every one of the Bible writers lived long before modern science existed. The writers of the Bible were authentic members of their own prescientific time and of their own prescientific culture, writing to prescientific people. The kinds of answers we find in Genesis are nonscientific because the Bible is not a science book. When Genesis is read with ancient eyes, we don't learn about modern science, but we *do* learn about God.

5

There Might Be a Time Machine in Your House

It was a glorious, fortuitous accident. Arno Penzias and Robert Wilson found what they weren't looking for but were smart enough to know they'd found something big.

It was 1964 and radio astronomers Penzias and Wilson wanted to analyze radio waves emitted by galaxy clusters. A twenty-foot horn-shaped telescope had been decommissioned from an early satellite communication system, and Penzias and Wilson were offered the telescope for their research.

There was one persistent problem, however. No matter where they aimed the telescope, they picked up odd static coming from all parts of the sky evenly, at all times. This mystery "noise" was microwave radiation. One by one, they ruled out every possible source of the noise.

Eventually, they were down to their final suspects—a couple of pigeons had taken up residence inside the antenna. In a final effort to solve the mystery, Penzias and Wilson re-homed the pigeons, then set out in their lab coats with scrub brushes to scour droppings off the equipment. Nevertheless, the buzz persisted. Microwave radiation, evenly spread throughout the sky, was still detected from every direction.

Meanwhile, Bob Dicke and his team at Princeton University were trying to find evidence of the "Big

Bang" explanation of the beginning of the universe. As the newborn universe expanded, Dicke predicted, the radiation given off would eventually cool and be detectable as microwaves. Dicke and his team looked for the radiation but had been unsuccessful in finding it.

On an unrelated phone call, Penzias decided to run the problem of the mystery static by Dicke. Penzias described the mysterious noise: evenly distributed microwaves, coming in at all times and from all parts of the sky. Dicke put down the phone, turned to his team and said, "Well boys, we've been scooped."[1]

Penzias and Wilson found the oldest radiation in the universe, and with it, evidence of the big bang birth of the universe. This radiation is called the cosmic microwave background (CMB) and it originated a mere 378,000 years after the big bang. For their discovery of the cosmic microwave background, Penzias and Wilson won the 1978 Nobel Prize for Physics, which they probably wouldn't have done if it had been pigeon poop.

In the days before cable and ten thousand TV channels, there was such a thing as an "in-between" channel, a dial setting on your analog set where all you could see was snowstorm-looking static. If you still have an analog TV in your home, you have a time machine. Set it to a static-y channel and invite your friends over to witness an event almost fourteen billion years in the past. One percent of the snowy static is cosmic background radiation being picked up by your television![2]

The Age of the Universe and the Earth

Genesis says nothing directly about the age of the earth and universe. However, adding up the genealogies in

Genesis and assuming seven literal 24-hour days of creation, young earth creationists date both the universe and the earth at about six thousand years old, but no more than ten thousand years old.

The most widely accepted science explanation for the beginning of the universe is the big bang theory. Simply, the theory states that the universe began as an incredibly tiny, incredibly hot, and incredibly dense single point, or *singularity*. From this one point, the universe inflated and expanded over 13.8 billion years.

In 1928, Edwin Hubble (for whom the telescope was named) discovered something amazing. Virtually all of the galaxies in the universe are moving away from us. Simply put, the universe is expanding. If the universe is expanding, it must have been smaller in the past. And since the universe is not infinitely large, there must have been a point in time when the expansion started from a single point—"The Big Bang."

We know the universe is expanding, but when did the expansion commence? How do we determine the age of the universe and the earth?

In the hard sciences as well as in the social sciences, the best conclusions are reached following a process called triangulation. If multiple ways of measuring something using different methods and processes converge at or near the same conclusion, we have confidence in the conclusion. We have triangulation.

In determining the age of the earth and the age of the universe, scientists use multiple, unrelated approaches and methods. Some approaches are very technical and rely on mathematics, others employ simple observational skills. All converge at the same conclusion: the earth and the universe are ancient.

How Old Is the Universe and How Do We Know?

If we know the speed of light, and we know how far a star is from earth, we can calculate the time it takes for light from the star to reach the earth. We begin with the distance between earth and the stars. We can actually measure the distances, but it's a little more math-y than using a cosmic meter stick. Conceptually it isn't hard at all, but if you want the trigonometry, to the *Google* for that!

Pick a spot on the wall in front of you. Close your right eye, now close your left eye. Your spot on the wall appears to move. You can also get the same effect by moving your body slightly to the left, then slightly to the right. As you move, note the objects directly in front of you and the objects farther away. The objects closest to you appear to move more than the objects farther away.

This effect is called a parallax:[3] far away objects appear to move less than do closer objects. As the earth moves around the sun, stars appear to shift. We can measure the angle of the star-shift in January, then measure the angle of shift again in July. With these measurements, we can construct a triangle. The diameter of the earth's orbit around the sun is the base of the triangle, and the angle opposite the base is how far the star appears to shift between January and July.

This is where your high school trigonometry comes in: measure the height of the triangle and you have the distance from the earth to the star.

The European Space Agency's Gaia telescope, launched into orbit around the sun in late 2013, is able to measure these angles with incredible accuracy. Gaia is set to give us the most detailed three-dimensional map of our galaxy to date.[4]

The Universal Speed Limit

The universe has a speed limit, and it is 186,000 miles per second: the speed of light.[5] Nothing travels faster than light.

The sun is ninety-three million miles away, so light leaving the sun has to travel ninety-three million miles before we can see it. If you do the calculations, you will find it takes eight minutes for light from the sun to reach the earth. When you bask in sunlight, you are actually looking eight minutes back in time. If the sun went dark, we would not know for eight minutes. Light from the nearest (non-sun) star takes 4.3 years to reach us. Light from the center of our Milky Way galaxy takes about 8,500 years to reach us. And when you observe the Andromeda galaxy, you see it as it looked 2.3 million years ago.

Other stars are unimaginably farther away, so far away it takes billions of years for the light from those distant stars to reach us. One way to estimate the age of the universe is to calculate how long it takes for the light we see from distant stars to reach earth. Considering the speed of light and the location of the most distant objects known at this point, the age of the universe is around 13.8 billion years.

The Universe Is Expanding

Have you ever noticed how the sound of a siren on an emergency vehicle changes from the time you first hear it to the time it rushes past you? As the siren approaches you, sound waves are bunched up together in front of the vehicle. "Bunched up" waves sound high pitched to your ears. As the siren passes you and travels away from you, sound waves are stretched. We hear stretched sound waves as lower-pitched. There is a name for this apparent change in the pitch (frequency) of sounds as they travel away from you: the Doppler effect.[6] Police radars use the Doppler effect to measure the speed of vehicles.

Light waves behave in the same way. When a star is moving toward the earth, light waves are all bunched up in front of the star, giving the light a higher frequency than normal. We see high frequency light as blue. Light waves from a star moving away from us is stretched out, giving the star a lower frequency. We see low frequency light as red. Light from stars moving toward us is termed *blue-shifted*; light from stars moving away from us is termed *red-shifted*.

Almost all of the light in the universe is red-shifted. And this: the galaxies farther away from us are more redshifted than the galaxies closer to us. The Doppler effect allows us to know how fast galaxies are traveling away from us.

Knowing the rate of expansion of the universe and the average distances between galaxies, we can calculate how long the universe has been expanding from the hot, dense singularity starting point until now. Using these calculations to push a cosmic rewind button, we know the singularity was about 13.77 billion years ago.[7]

How Old Is the Earth and How Do We Know?

Measuring the age of the earth can get very technical very quickly. It's easy to get lost in the chemistry and physics and easy to give up in the details. You need to know about those methods, but let's start with ways of measuring that don't require complex chemistry or physics—in fact, they do not even require complicated scientific instruments. Your eyes will do quite nicely.

Tree Rings

Each year, trees produce a new layer of wood under the bark. Depending on environmental conditions, a tree may form a double ring or even no ring in a year but cross-checking with multiple trees in the area allows for an accurate count. The age of a tree can be determined by visually observing and counting growth rings in the wood. Trees in the Ancient Bristlecone Pine Forest[8] in the Sierra Nevada are the oldest trees in the world. Living trees in this ancient forest are 4,800 to 5,000 years old.

But those guys are the youngsters.

Bristlecone pines grow in a cold, dry climate. Wood from long-dead trees, now fallen to the ground, can remain intact for thousands of years. Dead trees lying in the midst of the living forest are nearly twice as old—almost ten thousand years.

Lakebeds

Lakebeds accumulate sediments according to the season: mineral sediments in spring and pollens and plant materials in fall. Counting these accumulated layers (called *varves*) allows us to determine the age of a lake.

Lake Suigetsu in Japan has never been covered by glaciers and it is not fed by turbulent rivers, making it an ideal location for sediment sampling.[9] Sediment layers lie in pristine condition on the lakebed, relatively undisturbed since they sank to the bottom. Cores taken from the bed of Lake Suigetsu establish its age to be at least sixty thousand years old.

Ice Cores

Ice sheets and glaciers form from years and years of accumulating snowfall. Each season's snowfall tells a story about what the earth was like in the year the snow fell. Particles in the air such as dust, pollen, and ash are trapped in the ice, and even gases can be trapped in the form of bubbles. Each layer is unique to the year in which it was laid down, with slightly different combinations of particulates and gases. These unique layers can be counted and used to determine the age of the ice, just like the rings in a tree.[10]

Researchers drill into ice around the world in order to study climate histories. Ice cores are removed, and the layers counted. Ice cores from Greenland are a mere 130,000 years old. The oldest ice cores to date are in East Antarctica at 800,000 years old.

Anyone can count rings of new wood and layers of sediments and pollens—no special skills or instruments needed. Using the simple skills of visual observation and counting, tree rings, lakebeds and ice cores place the earth well past the 6,000-year age proposed by young earth creationists.

Radiometric Clocks

Observational data allows us to date the earth to almost one million years, but other methods allow us to date rocks back into deep, deep time.

In the nucleus of an atom are particles called protons and neutrons. Usually, all is happy and stable in the atomic world. Sometimes, however, an atom is unstable and will "throw off" some of the particles in its nucleus. This is called radioactive decay.[11]

As a radioactive element throws off particles, it may transform into a different, more stable element. The time it takes for half of a radioactive element to decay into a more stable element is called the *half-life* of the radioactive element. Some radioactive elements decay very quickly and have half-lives of seconds or days. Other radioactive elements decay very slowly and have far longer half-lives. Knowing the half-life of a radioactive element allows us to determine the age of the rock in which it is found.

For example, atoms of uranium-238 are unstable. Uranium-238 will throw off particles until it has changed into a more stable element, lead-206. The half-life of uranium-238 is enormously long—4.5 billion years! When we want to date a rock containing uranium-238 and lead-206, we measure the ratio of uranium-238 to lead-206. Knowing this ratio allows us to determine the age of the rock using a "radioactive clock."

The oldest rocks on earth (to this point) are found in western Australia's Jack Hills region. Using a uranium-lead clock, crystals in the rock were found to be almost 4.4 billion years old.[12] Interestingly, moon rocks are dated at 4.5 billion years using the same method. Even allowing for small errors in calculations, the earth is ancient, far exceeding a timeframe of six thousand to ten thousand years. Other radioactive clocks with slightly shorter half-lives are more limited in dating power, yet still point to an earth more than a billion years old.

Magnetism

Deep inside the earth, incredibly hot liquid iron sloshes around the outer core. The sloshing movement of molten iron creates a magnetic field, like a giant imaginary bar magnet inside the earth. And like an actual bar magnet, this imaginary magnet has north and south poles, giving the earth its magnetic field.

The molten iron inside the earth is not static, it is constantly moving and swirling. In all the swishing and swirling, some iron atoms get flipped in the opposite direction from the iron atoms around them. When enough atoms flip, the imaginary bar magnet flips, and the magnetic north becomes the magnetic south.

The earth's magnetic poles have flipped hundreds of times in earth's history. How do we know? The earth's magnetic field determines the magnetism of lava as it flows over the ocean floor on either side of the Mid-Atlantic Rift, where the North American and European continental plates are slowly separating. When the lava solidifies, a record of the past magnetism of the earth is set in stone. The lava record tells us that about every 200,000 to 300,000 years, the poles flip. Sometimes the timeframe is longer; it's been 780,000 years since the last flip. All told, the earth's magnetic field is at least 3.5 billion years old, about the same age as the earliest fossils we've found so far.[13]

Why a Young Earth?

Forty percent of Americans are young earth creationists.[14] The percentage jumps to 68 percent among weekly churchgoers. A belief in a young earth is always coupled with nonbelief in evolution. You just do not find young earth proponents who accept evolution. Why might this be so?

The most obvious reason is a commitment to a literal Genesis. Literal 24-hour days in a literal seven-day week, coupled with a literal accounting of Genesis genealogies requires a young earth.

The greater appeal to a young earth, however, comes as opposition to evolution.

Even the most ardent young earth creationists concede that species change—just a bit—in order to meet environmental challenges. Although creationists (both young and old) accept small adaptive changes ("microevolution"), these small changes never result in a new species: finches remain birds despite changes in beaks, dogs remain dogs despite variations in appearance, horses remain horses despite variations in size, and so on. Creationists (whether young earth or old earth) do not accept "macroevolution" in which a species gives rise to other new and different species.

Enter Charles Darwin and his game-changing concept of deep time. To tiny, incremental, imperceptible changes, Darwin added time. Lots and lots of time. Millions and millions of years. Deep, deep time.

Species do not change suddenly in big flashy events. A modern bird never hatched from a dinosaur egg. Species give rise to new species when small changes accumulate over deep, deep time. Microevolution events, over time, result in macroevolution.

In 2005, Mary Schweitzer, former graduate student of famed dinosaur hunter Jack Horner, astounded the paleontology world. Using modern cell biology techniques, Schweitzer found soft tissue in a dinosaur fossil—a 68-million-year-old dinosaur fossil. Predictably, paleontologists were skeptical. Soft tissue still intact after all this time? Soft tissue surviving despite fossilization? How is that possible?

Creationists, however, were ecstatic. Here, at long last, is undeniable evidence for a young earth. So much for your millions-of-years fossil formation. So much for your billions-of-years rock dating. Here is (almost) living proof: soft, pliable tissue from a *Tyrannosaurus rex*, buried four thousand years ago in Noah's flood. For creationists, dinosaur soft tissue is a strike against modern radiometric rock dating, but the death blow is to evolution: "if the rocks and fossils are not millions of years old, evolutionary theory is finished."[15]

Since Schweitzer's initial discovery, more soft tissues have been extracted by Schweitzer and other researchers. Most importantly, Schweitzer and others have demonstrated several processes capable of preserving soft tissues in ancient fossils.[16] While Dr. Schweitzer is a hero to creationists for her initial discovery of soft tissue, creationists thoroughly dismiss all of her work demonstrating soft tissue preservation in fossils. A self-described "complete and total Christian," no one is more dismayed by the hijacking of her work by creationists than Schweitzer herself.[17]

The Cost of a Young Earth

Observable evidence in trees, lakes, and ice cores ages the earth at almost one million years old. Multiple lines of chemical and physical evidence age the earth at about 4.5 billion years old and the universe at about 13.8 billion years old. If the earth and universe are actually young (six thousand to ten thousand years old), one of two things must be true:

1. Evidence is being misunderstood and misinterpreted. If scientists use the Bible to interpret the data, they will conclude that the earth and universe are young.
2. Just as Adam and Eve were created as adults, the earth and universe were created "full grown." Light, rocks, ice, trees, and lakebeds were all created with the appearance of age.

What is the cost of a young earth and universe? Insisting on a young universe and a young earth ignores not just one line of evidence, but many lines of evidence. Insisting on a young universe and a young earth not only dismisses somewhat complicated chemistry and physics, but also dismisses clearly seen observational evidence in trees, lakes, and ice.

Insisting on a young universe and a young earth ignores not just one line of evidence, but many lines of evidence.

Insisting on a young universe and a young earth requires dismissing chemical and physical principles routinely used in modern science and technology for purposes other than determining age. Is it reasonable to trust the physics, chemistry, and mathematics in aeronautics and space travel and in every field of modern engineering, but disbelieve the exact same science when it tells us the age of the earth and universe?

Insisting on a young earth requires dismissing not only science evidence, but also evidence from archaeology. We have ten thousand years of evidence of human settlements with domestication of plants and animals in the Middle East. We have 12,000-year-old pottery from Japan. We have 35,000-year-old art in Europe. Observable human history quickly bypasses the 6,000-year mark.[18]

Insisting on a young earth discredits the fossil record, modern physical sciences, and archaeology, all in one fell swoop.

6

It's Raining, It's Pouring, the Canyon Is Forming: Noah's Flood Explains It All

It must have been a sight.

London, 1872. Tucked away in the depths of the British Museum, curator George Smith suddenly jumps up from his work and runs around the room, whooping and hollering.

And—to the astonishment of his colleagues—he proceeds to strip down to his underpants.

George Smith was translating an ancient Assyrian tablet, dated to the seventh century BC. He came upon a story in the ancient tablet about a man and a devastating flood: forewarned by the gods, the man built a big boat and filled it with animals and his family. Later in the story, he sent out a dove and raven to check for dry land.

All this in a story predating the book of Genesis.

The narrative Smith discovered came to be known as the *Epic of Gilgamesh*, after the story's boat-building hero. It was an extraordinary find, and it remains one of the oldest works of literature in the world.[1]

A Round Ark

In 2014, a newly deciphered 4,000-year-old Babylonian cuneiform tablet went on display in the British Museum.[2] The tablet looked like a cell phone-sized shredded wheat biscuit. It was given to the museum by a man whose father had acquired it in the Middle East after World War II.

Irving Finkle, translator of the new tablet, discovered it also contained a flood story, significantly predating both the Gilgamesh story and the Genesis account. Finkle was delighted to find the animal-gathering instructions were more technical ("two by two") than the instructions found in Smith's story. The biggest splash, however, was the detailed description of the boat: giant, made of rope, and *round.*

Finkle identified the boat design as an oversized *coracle,* vessels used in ancient Iraq as water taxis. Round coracles are perfectly designed to bob along on rushing floodwaters.

Geophysicists William Ryan and Walter Pitman documented evidence of an enormously catastrophic flood in the Middle East about 7,500 years ago.[3] Evidence suggests that as sea levels rose following an ice age, the Mediterranean Sea overflowed and deluged the Black Sea basin. According to estimates, water could have rushed through this channel with forces greater than Niagara Falls, with water levels rising six inches per day.

> *It is not a surprise, then, to find flood stories in the collective memories of ancient cultures in that part of the world, Israel included.*

There are many other records of catastrophic floods in Mesopotamia,[4] dating back to 2900 BC. Observable flood deposits in the area coincide with these accounts. It is not a surprise, then, to find flood stories in the collective memories of ancient cultures in that part of the world, Israel included.

Flood Geology

Discoveries of flood stories more ancient than Genesis make many Christians uneasy. If your biblical interpretation requires the Noah story to be the original flood story, other more ancient stories might be unsettling. Things get more complicated if your biblical interpretation requires a literal worldwide flood in order to explain geology and fossils.

The modern study of geology and fossils began in the late eighteenth century. These discoveries were not made by professional scientists, but by independently wealthy gentleman-naturalists and others who were not so wealthy. Usually, these laymen-scientists were devoutly religious.

Well into the twentieth century, many protestants, including fundamentalists, accepted an ancient earth with "progressive" creationism. As the nineteenth century progressed, however, some Christians were increasingly uncomfortable with the ever-expanding geologic evidence of change over time, especially in light of Darwin's recently published work. The first notable attempt to reconcile Genesis with geologic evidence came in 1902 from a Seventh-Day Adventist schoolteacher, George Macready Price. Inspired by visions reported by Seventh-Day Adventist founder Ellen G. White, Price coined the phrase "flood geology." According to Price, the varied geologic features of the earth along with deposited fossils are explained by Noah's worldwide flood, as described in the book of Genesis.

Price's work was largely ignored for most of the twentieth century, even among creationists. In 1961, seminarian John Whitcomb and engineer Henry Morris published *The Genesis Flood*,[5] rehashing many of Price's ideas and adding their own unique spin regarding fossil deposition. *The Genesis Flood* popularized "flood geology" in broader evangelical and fundamentalist circles and established the modern young earth creationism movement.[6] Several creationist societies, think tanks, and publishing houses were launched in the wake of *The Genesis Flood*'s publication.

Flood geology still lives on in the twenty-first century.[7] Many in the young earth camp attribute earth's complex geologic history to Noah's flood. From continental drift, to the Grand Canyon, to fossil layers, to the demise of the dinosaurs—a literal, global flood is credited as the cause.

The Grand Canyon is particularly a favorite case-in-point for flood geology proponents. According to the flood geologists, the Grand Canyon formed rapidly as the result of turbulent floodwaters, then was further carved out by the Colorado River over several thousand years. Flood geologists believe fossil-bearing sediments were laid down in the canyon, as well as over the rest of the earth, in the one-year timeframe of Noah's flood.

In the widely released film (2017) *Is Genesis History?*,[8] the Grand Canyon is front and center. It is a beautiful movie—lots of on-location shots from the Grand Canyon, with emphasis on its origin via a catastrophic world flood. Host Del Tackett (Focus on the Family) summarizes the complex geology of the Grand Canyon with a twist of Dobzhansky's famous quote ("nothing in biology makes sense except in light of evolution"):[9] "Nothing in the world makes sense except in light of Genesis."[10]

The film features seventeen speakers from various fields, including geologists, biologists, an engineer, and pastors. Each speaker dismisses, out-of-hand and without elaboration, any references to a canyon older than four thousand years. All speakers agree that the canyon was formed suddenly and catastrophically by raging and then receding flood waters. At several points, evidence contradicting flood formation is mentioned, but each mention is dismissed immediately. Any conflict between science and the film makers' interpretation of Genesis is awarded to Genesis. One speaker cites the evidence regarding the age of the earth as "claims of Scripture and my own experience."

Canyon Evidence

What does the geologic evidence indicate regarding the formation of the canyon?

For many years, I taught a "how to teach science" course to elementary education majors in university. In our soil lesson, we filled a glass jar with a half soil, half water mixture. We then shook the jar vigorously until it looked like chocolate milk. We let the mixture settle for a few hours and this is what we saw:

Heavier rocks settle out first, followed by finer sand, followed by finer silt, then clay, the finest of all.

In geology, this phenomenon is called *fining upward*. When flood waters recede, we observe a "fining upward" sequence in the layers of soil laid down by the floodwaters—coarse layers at the bottom, getting finer and finer toward the top.

If a single catastrophic flood—a single surge of rushing water—was responsible for carving the geologic features of the earth such as the Grand Canyon, what would we expect to see? What would we expect to see after the surge-event retreated and the soil settled down? We would expect to see coarse layers at the bottom of the canyon, with finer and finer layers to the top. Last to be laid down would be a single layer of mud.

But that is not what we see. Deposition in the Grand Canyon is a series of alternating layers—fine, coarse, fine, coarse, fine, coarse, and so on. Some of the alternating layers are larger than others. And the upper part of the canyon where we would expect to see a layer of mud following a great flood, shows no such thing. The upper part of the canyon is not mudstone, but layers of shale, sandstone, and limestone in formations never before seen in in flood deposits. The rocks in the Grand Canyon bear witness to multiple episodes of deposition with intervening times of erosion over vast amounts of time.

Furthermore, not all sediments in the canyon are laid down in flat, horizontal layers. In many areas, we see buckled, fractured, and faulted layers, testifying to a long history of different kinds of forces pushing, pulling, and stressing the rock. Because flood geologists restrict the timeframe for canyon formation to the year or so duration of Noah's flood, they reject this evidence. Instead, flood geologists attribute rock deformations to the folding of soft sediments freshly laid down by the flood.

If rock deformations are the result of folded soft sediments as flood geologists claim, what would we expect to see in the canyon today? Freshly deposited sediments are mostly water. Layers of freshly deposited flood sediments stay separated as long as everything is calm and still. However, as soon as folding begins, the watery sediment layers lose their consistency as they blob, flow, and mix with each other. Actually, there are isolated incidences where soft sediment deformation has occurred, and it has a distinctive appearance.[11] However, we do not find this type of rock in the Grand Canyon.

We also find preserved mud cracks, raindrop prints, and water current ripples in Canyon rock. Apparently, sediments were intermittently exposed to long periods of air and/or shallow water, impossible if the canyon was formed by a year of surging floodwaters.

Now, throw multiple layers of limestone into the mix, and global flood explanations for the Grand Canyon are in serious trouble. Limestone commonly forms in shallow marine waters where soft-bodied animals lived and died, leaving behind shell fragments and skeletal remains. The largest cliff faces in the Grand Canyon are made of limestone. There are innumerable examples across the globe of limestone forming in shallow marine waters. Limestone does not form in turbulent floodwaters.

Beyond the Canyon

Flood geology explanations extend well beyond the Grand Canyon. Flood geology says that forty days of upheaval and rain, followed by several months of receding waters carved the earth's surface and laid down the entire sedimentary record, including fossils.

What does the geologic evidence say?

Fossilized Deserts

Stretching east from the United Kingdom, across Europe and into Russia, and extending west into Greenland and the northeastern seaboard of North America, is a geologic formation called the Old Red Sandstone. The appearance of the formation is unmistakable: Old Red Sandstone formation is composed of fossilized sand dunes. In North America, we find beautiful petrified dunes in Zion National Park, Red Cliffs Desert Reserve, Arches National Park, and the Coconino Sandstone of the Grand Canyon.[12]

Ancient sand dunes, turned to stone, are the result of a combination of mechanisms: pressure from an increasing mass of newly laid down sand combined with the cementing effect of dissolved minerals in groundwater. Some dunes preserve evidence of ancient inhabitants—invertebrate tracks and burrows and reptilian footprints. The Old Red formation features hundreds of square miles of dunes containing cross-bedding layers and sand-blasted pebbles, features commonly found in modern desert sand dunes, and in no other type of sediment.

What would we expect to see if the sand dunes formed underwater, and turbulent water at that? Would we expect to see preserved tracks of tiny creatures? Would we expect to see dune banding and blasted pebbles unique to wind formation? Everything we see in fossilized sand dunes speaks to their formation in a dry desert environment.

Salt Deposits

There are many massive salt beds on earth, some thousands of feet thick. We know how salt beds form—we can observe actively growing salt beds such as those in the Bonneville Salt Flats in Utah.[13] Salt beds form when salty water evaporates.

Flood geologists attribute the massive salt beds found throughout the world to water evaporation following Noah's flood. There is an inconvenient problem with this explanation: the salt beds are covered with thousands of feet of sediment, also said to be left by the flood. Here's the dilemma: water from the flood must evaporate enough in a year's time to produce vast amounts of salt, but the water must still be massive enough in volume to produce thousands of feet of sediment on top of the salt.

Fossils

> Now, if there really was a global flood, what would the evidence be? Well, there'd be billions of dead things, buried in rock layers, laid down by water, all over the earth. And, you know, that's exactly what we see![14]

George Macready Price was the first to suggest a global flood was responsible for fossil layers. Whitcomb and Morris and their *Genesis Flood* took the idea and ran with it.

If a global flood was responsible for the layers of fossils we observe today, what would those layers look like? Waters turbulent enough to rip up and move continents would churn all unfortunate non-ark animals as well as the bodies of previously deceased animals into one big animal-soupy concoction. There would be plants, too—all kinds—from simple algae to the tallest trees and everything in between, in a chaotic chopped salad of vegetation. There would be no orderliness to the fossil record; instead, we'd have the paleontological equivalent of a scoop of rocky road ice cream.

But that is not what fossil layers look like. The oldest rocks have only simple mats of bacteria. In very old rocks, we find fossils of soft-bodied animals and the first evidence of plants. From oldest to newest, we see an unfolding of life: trilobites, fish, the first four-limbed animals, and eventually, dinosaurs. Later still, we find mammoths and other mammals . . . and humans. Flowering plants are found in the newest rocks, but not in older rocks.

Why Is a Global Flood So Important in Creationism?

The first and most obvious answer is a commitment to biblical literalism. For an inerrantist, Noah's flood and the ark are literal history, in both detail and timeframe. The commitment, however, is deeper than surface literalism.

From the top of Everest to the depths of the Marianas Trench, the earth's crust testifies to the action of dynamic forces over eons of time. The crust is by far the thinnest layer of the planet, less than one-half of one percent of the earth's total mass. Yet, what a history the crust reveals! Continental plates float, move, and crash into each other. Mountains thrust up in the wake of continental collisions. Volcanoes erupt where openings in the crust allow the movement of molten lava and gases. Tremendous heat and pressure changes rock from one type to another. Rocks are churned from the deepest layers to the surface. Glaciers glide on the crust's surface, gouging out lakebeds. And, of course, magnificent features like the Grand Canyon take shape, cut by water, weathering, and erosion.

Time is a big problem for young earth creationism. The problem of vast amounts of time needed for a thrust up, moved, and gouged-out crust is answered by *catastrophism.*[15] Instead of a testimony to eons of time, the features of the earth's crust are a result of a catastrophic event, specifically, Noah's global flood. Catastrophism attributes not only the Grand Canyon, but almost all geologic features to a global flood.

Modern catastrophism goes far beyond torrents of rain and universal flooding. Subterranean bodies of water burst through the earth's crust. Continents are ripped apart and then collide. Entire mountains are slammed about. The ocean floor is ripped to shreds. And all of this destruction occurs within forty days and nights, solving the problem of the geologic timeframe.

Noah's flood is important to creationism for reasons of biblical literalism and for explaining the formation of geologic features in a short period of time. However, the significance of Noah's flood to creationism runs much deeper than a literal Genesis and a catastrophic Grand Canyon formation. Noah's flood solves the biggest problem of all: evolution.

If Noah's flood explains the vast fossil record, fossils were not formed through eons of deep time. If fossils represent a mass all-at-once death of living creatures, we cannot draw evolutionary inferences from the fossil record. If Noah's flood produced the fossils, we have no record in stone of change over time. Noah's flood is central to an anti-evolution narrative:

> Instead of representing the evolution of life over many ages, the fossils really speak of the destruction of life in one age, with their actual local "sequences" having been determined by the ecological communities in which they were living at the time of burial.[16]

If the flood produced the fossils, we have no evidence for evolution.

In a quirk of fate, Ken Ham's Kentucky Ark Encounter sits on top of a famous Ordovician geologic formation known as the Cincinnati Arch.[17] The Cincinnati Arch is well known for its hundreds of finely laminated layers of shales and limestones, brimming with delicate fossils preserved in undisturbed life positions, gently buried in fine silts and clay. Fossils of corals show many years of growth bands, evidence they grew in one location over long periods of time. The panorama of fossils on which the Ark museum sits could not possibly be the result of catastrophic and turbulent flood waters.

Yet, flood explanations for fossils persist. On to the next chapter for more on fossils and flooding!

7

The Flood and the Fossil Record

Victorians loved their curiosities. They were particularly fond of taxidermic animals. In fact, many of the displays we see in natural history museums are artifacts of pre-PETA days when animals were stuffed and collected as a hobby. Amateur taxidermist Walter Potter,[1] however, took the cake. In a peculiar blending of Victorian whimsy and Victorian fascination with death, Potter created tableaus with his handiwork: bunnies hard at work in a schoolhouse, kittens at a tea party, card-playing squirrels, and my favorite, a cat wedding.

Victorians also loved to collect other objects from nature. They called these prizes "curiosities." Victorians often displayed their curiosities in a cabinet (the "curio cabinet") or in a special room in their home.

Lyme Regis, a seaside resort in Dorset county in England was a favorite vacation spot for those with enough money to take a holiday on the coast. In addition to the beauty of limestone and shale cliffs, the area was noted for an abundance of fossils. Victorians did not know how to explain these fossil curiosities, so they concocted their own explanations:

- Fossilized vertebrae were called "verteberries" or "crocodile teeth"

- Beautiful ammonites (an extinct mollusk) were called "snakestones" or "serpent stones"
- "Devil's fingers" or "St. Peter's fingers" are extinct mollusks similar to modern squids.

"Angels' wings," "devil's toenails," and more—the Victorians didn't know what they were, but they loved the mystery and they loved to collect them.

The Bone Girl

Richard Anning was a poor cabinet maker in Lyme Regis in the early nineteenth century. In the endless struggle to keep his family fed, he collected fossils to sell. He set up a little table in front of his shop and sold his curiosities, small fossils and seashells, to the vacationers. Fossil hunting and extracting in the cliffs could be dangerous work, but Anning's two children often accompanied him as he searched. He even made his little daughter Mary a fossil extractor of her very own. When Mary was only eleven, her father died from consumption following a fall from a cliff. The little family edged closer to destitution.

Not long after their father's death, Mary's brother noticed a skull with a ring of bony plates around the eye socket—they thought it was a crocodile—but in England? A year later, twelve-year-old Mary returned to the site and found the rest of the creature's skeleton on a cliff high above where the head was found. Young Mary led a group of men to dig out the skeleton—an almost perfectly preserved seventeen-feet-long reptile. It was *not* a crocodile— it was a 200-million-year-old ichthyosaur, a marine reptile with paddle-like limbs and a streamlined body.

Scientists in the fledgling fields of geology and paleontology often came to Lyme Regis, but with the discovery of the pristine ichthyosaur fossil, several stars in the fields specifically sought out the teenaged Mary. There were many more discoveries by Mary over the years: long-necked plesiosaurs (including the first two specimens ever found), more ichthyosaurs, a squid-like cephalopod, an ancient starfish, and ancient fish. She even discovered the first pterosaur (a flying reptile) found in Britain.

During the Jurassic geologic period (about 206–144 million years ago), the Lyme Regis area was submerged in a vast shallow sea teaming with life, a banquet for large carnivorous marine reptiles and for the pterosaurs living along the shoreline. Mary, who travelled out of Lyme Regis only once in her life, was a smart woman in just the right place.

But in class-conscious Victorian England, she was poor. She was a woman, and an unmarried woman at that. She had little formal education. Because she sold her finds to museums and collectors, she was not considered a scientist; she was "in trade." Mary was religious and deeply faith-filled, but she belonged for most of her life to a "Dissenters" church—not the respectable Church of England. She never married, but she supported her mother and was devoted to her little dog, her fossil-hunting companion. To her sorrow, the little dog was killed in a rockslide which narrowly missed Mary.

Although Mary had little formal schooling, she was far from uneducated. She read and educated herself in her field—particularly comparative anatomy. She was respected by the early leaders in the field of paleontology, but they never named the finds for Mary. In 1835, the British Association for the Advancement of Science awarded her a modest lifetime annuity in recognition of her work—remarkable for the time, since women were not expected to be highly educated, much less scientists. Mary died twelve years later at age forty-seven from breast cancer.

There is a breathtaking sun-drenched gallery hall with high, large windows in the Natural History Museum in London. Both sides of the hall are filled, floor to ceiling, with fossilized marine reptiles found in England: ichthyosaurs, plesiosaurs, and more. On the plaques of some the finest fossils in the gallery, you'll see the name "Mary Anning," time and time again. Some days, a museum docent dresses as Mary and interacts with museum visitors: "the greatest fossilist the world ever knew."[2]

Have you heard of Mary Anning? Probably not. But I'm sure you've heard this:

> She sells seashells on the seashore,
> The shells she sells are seashells, I'm sure,
> For if she sells seashells on the seashore,
> Then I'm sure she sells seashore shells.

This children's tongue twister was written about Mary Anning!

Mary stood at a much more profound crossroad of science than she could have fathomed. Scientists of Mary Anning's day could not comprehend deep time—time measured in billions of years, not thousands. Geology was a new field; scientists were just beginning to understand the forces that shaped the planet. Paleontology was even newer: when Mary was born, dinosaurs had not yet been found. Dinosaurs were not identified as a group and named *Dinosauria* until just a few years before Mary's death.

A few naturalists (as biologists were then called) considered the possibility that life had changed over time, but there was no way to frame such changes given the age the earth was assumed to be. Mary lived, worked, and died before Charles Darwin burst onto the scene. Although she and Charles Darwin were contemporaries, Darwin did not publish *On the Origin of Species* until ten years after Mary's death.

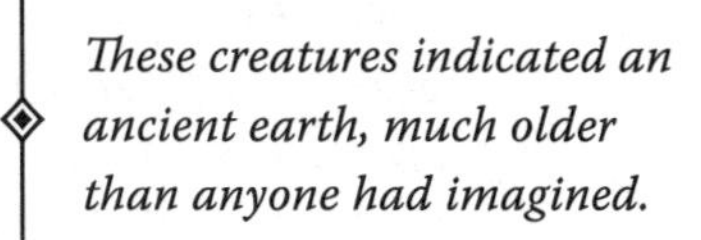
These creatures indicated an ancient earth, much older than anyone had imagined.

Pretty little seashells on a table in front of a curiosity shop threatened no one. Small fossilized marine animals were curious—but weren't terribly threatening. Victorians loved them and the mystery: were they medicinal? Were they sinners turned to stone? But giant fossilized marine reptiles buried deep in the rock were threatening.

These creatures indicated an ancient earth, much older than anyone had imagined. And apparently, life on earth had been very, very different in the past. As Mary found specimen after specimen, the challenges to existing beliefs about creation and the meaning of the Genesis stories grew stronger. It was unavoidable: time was unfathomably deep. Life on earth had changed. Victorians loved the fossil curiosities but could no longer ignore the implications.

The Rocks Tell a Story

If your biblical interpretation requires special creation of each kind of living thing six to ten thousand years ago in six literal days, you might struggle with the implications of the fossil record. Even if you are an old-earth creationist and believe special creation took place over a long period of time, you might struggle with the sequence of life revealed in the fossil record. And if you believe fossil layers are the result of a catastrophic global flood, the science evidence may be especially challenging.

Both schools of creationism, young and old, acknowledge that the fossil record challenges the order of creation in Genesis. Problems arise when questions like "why are there no human fossils in layers containing dinosaur fossils?" are posed. More complex problems arise when attempting to explain, for example, why the oldest rock layers lack fossils or only show evidence of simple microorganisms.

Regardless of the problem, creationists reject outright what they term "naturalistic earth history," the practice of aging fossils by aging the rocks in which they are found. The creation order in Genesis is in conflict with naturalistic history, and the conflict is insurmountable.

Most creationists resolve the conflict by utilizing a Bible-based geological framework of earth history, to the exclusion of all other frameworks. There is some disagreement among creationists as to what extent a "Bible-based" framework can take into account "naturalistic" observations, but regardless, the framework starts with Genesis. Paul F. Taylor puts it this way: "(we) explain observed facts through the 'eyeglasses' of the Bible. The Bible is inspired, but our scientific models are not."[3]

How, then, do creationists explain the fossil record using Bible "eyeglasses"? With the caveat that a few fossils were deposited pre-flood and others were laid down post-flood, creationists explain the bulk of the evidence we find in the fossil record as the result of Noah's global flood. *The Genesis Flood* (1961)[4] introduced the idea of hydraulic sorting by the global flood waters: heavy shells of mollusks buried first, followed by fish, followed by "advanced" animals like amphibians, reptiles, and dinosaurs, and finally the most advanced of all, the mammals, who climbed to the highest points above the flood waters trying to save themselves. The real winners were the birds who flew to safety until their demise ultimately arrived.

Twenty-first-century creationists extend the flood sorting explanation to include animal behavior and intelligence in addition to body weight and mobility. Ocean invertebrates appear in the earliest layers of fossils because they are heavy and slow. Fish, being more mobile, appear in layers of rock above marine invertebrates. Other animals are found in layers according to how "flexible" they are in behavior: amphibians, being the least intelligent are first, followed by reptiles, birds, and finally mammals. The most intelligent creatures find their way to high ground, and there they eventually die. For this reason, creationists are not surprised to find intelligent humans in the newest, or "top-most" layers of rock.[5]

Does a "sorting" explanation for fossil layers, either hydraulic or behavioral, hold water? Do we actually find neatly sorted out layers of animals, from heaviest and dumbest at the bottom to lightest, fastest, and smartest at the top?

It is true—organisms appear in the fossil record in a specific pattern. In the oldest rocks bearing fossils we see cyanobacteria—simple single-celled organisms without a cell nucleus. In the next layer up, we see single-celled organisms with a nucleus. Moving forward in time through younger and younger rocks, we see the first multicellular organisms, then we see organisms with heads, then organisms with four limbs, and in some of the youngest rocks, we see the first humans.

But this is important: while we see *first appearances* of increasingly more and more complex body forms in the fossil record, the least complex don't disappear. Once mollusks appear in the fossil record, their descendants continue to appear in subsequent layers. Once fish and four-limbed animals and humans appear in the fossil record, they continue to appear in subsequent layers.

What we *do not* see in the fossil record is a representation of all kinds of life in each layer of rock. We don't see animals with legs in the layers where fish first appear. We don't see humans in the layers of rock where multicellular organisms first appear or where four-limbed animals first appear or anywhere close to where we find dinosaurs. Is this what you would expect to find in flood-deposited soils from a catastrophic event that ripped continents apart and killed every living thing on earth, *at the same time*?

One of the richest vertebrate fossil-bearing areas in the world is found in Badlands National Park.[6] Here, many layers of rock spanning long periods of geologic time are exposed. At the base of the sequence we find mollusks: ammonites and clams. So far so good for the "heavy and slow and not-so-smart" explanation. In the next layer up, we find some truly impressive mammals, including the huge rhinoceros-like brontotheres. Above this layer, we find more large and swift mammals (deer-like animals and antelope-like animals) as well as an assemblage of small mammals: squirrels, rabbits, other small rodents, and tiny three-toed horses.

Here, the smarter and swifter idea begins to fall apart.[7] Apparently, the small mammals were swifter than the long-legged brontotheres in reaching the higher ground of safety. But most astonishing of all, the predominant animal group found in the uppermost formations are tortoises! Faster than the rodents, faster than the swift long-legged mammals, faster even than the hares—Aesop would be proud.

Recall, Mary Anning found ammonites (mollusks with large, heavy shells), huge and swift predators like plesiosaurs and ichthyosaurs, fish, starfish, and *flying* reptiles (pterosaurs), all in the Lyme Regis formation. If the "heavy and slow and not-so-smart" were the first to drown while the smarter and faster escaped until the last minute, is this what we would expect to see in the Lyme Regis and in the Badlands and in multiple other formations on the planet?

And what about the smartest and most agile of all—the humans? Human fossils are only found in the newest layers of rock. Creationists acknowledge this fact: being fast as well as the smartest, humans would be the last to die. Of the entire world human population (minus the ark eight), was no one slow? Or elderly? Or injured? Or sleeping? Was everyone a speedy adult? No children? Without exception, no humans appear in the fossil record until the newest layers. Not a single one.

Animals would not have been the only living things impacted by a year-long, worldwide flood. Where do plants fit into the picture? Just as animals appear in the fossil record in a recognized pattern, the fossil record for plants reveals a consistent pattern.

By far, the most abundant plants on earth today are the flowering and fruiting plants, the angiosperms. Angiosperms are wildly diverse; they include trees, shrubs, grasses, cacti, and food crops. Yet as abundant as angiosperms are, they are the new kids on the block. Angiosperms first appear in the fossil record in fairly young rocks, geologically speaking. Although we have a few fossils showing early angiosperm features (pollen grains, flower-like structures) in older rocks, identifiable angiosperms do not appear in abundance until about 65 to 100 million years ago.

Before plants appear in the fossil record, we find only algae. The first land plants to appear (445 million years ago) are simple plants—the mosses and liverworts, followed by vascular plants (430 million years ago). Over the next 200 to 300 million years, vascular plants diversified in body forms, modes of reproduction, and environments. Simple upright plants were the first, followed by ferns and horsetails, some as large as trees. The time of the dinosaurs was dominated by non-flowering seed plants—plants with seeds exposed on cones, like pine trees. And finally, last to the party, flowering plants appear.

What kind of plant layering would we expect to see following a worldwide flood with turbulent waters covering all exposed land? We would expect a mishmash of plant materials—leaves, stems, flowers, and roots, literally pulled apart limb from limb. We would expect to find few if any intact plants. We certainly would *not* expect to see a complete absence of plants in the oldest layers and an unfolding story of plants increasing in complexity and diversity in subsequent layers.

Genesis specifically says trees bearing fruit were created on the third day, yet these fossils are only found in the newest, uppermost layers. Creationist literature rarely addresses the plant fossil record or land plant survival after oceans flooded and covered the surface of the earth with saltwater. Compared to the "sorting out" explanation for animal fossils, little is said about the "sorting" of plants. I like how Kenneth Miller puts it: "Plants are good at many things, but running to high ground during a flood is not one of them."[8]

Kangaroos in Kayaks

What about the animals that did not drown in the flood, the pairs saved on the ark? What does a literal reading of Genesis require? Only the lucky few saved on the ark repopulated the earth. From a mountaintop in the Middle East, each pair would need to navigate somehow to the far reaches of the planet—a daunting re-homing project. Dispersal of animals around the world following Noah's flood is a logistical challenge.

Australia is obviously a particular dilemma. It is remote and isolated from the other continents and surrounded on all sides by the Indian Ocean. On top of that, Australia is home to a zany collection of mammals seen nowhere else on earth.

All living mammals on earth belong to one of three major groups. By far the most abundant, widespread, and well-known of the three groups are the placental mammals. Placental babies complete their embryonic development within the mother's uterus. Like all mammals, placental moms produce milk for their babies.

Marsupial mammals also have a placenta, but marsupials are born extremely early in their embryonic development. After they are born, tiny underdeveloped marsupials crawl into a marsupium, or pouch. There, they latch onto a nipple, nurse, and complete their development. Familiar marsupials are kangaroos and koalas.

Monotremes are the third mammal group and are by far the smallest group and the oddest. Monotremes lay eggs! Monotremes also secrete milk, but they do not have nipples; babies lick the milk from their mother's fur. The duck-billed platypus and the echidna are monotremes.

Australia is dominated by marsupials. In addition to kangaroos and koalas, Australia is home to wombats, sugar gliders, Tasmanian devils, and hundreds of other marsupial species. Only one type of marsupial is found anywhere outside of Australia: the opossums of South and North America. What's more, Australia is the *only* home for monotremes. What Australia lacks are the mammals dominating the rest of the planet: the placentals. The only native placental mammals found in Australia are ones that flew there (bats) or swam there (sea lions, seals, dolphins).

Australian mammals pose a difficult problem for creationists. Creationists do not claim Noah took two of every possible animal on board, but instead took pairs of generic "kinds": a dog "kind," a cattle "kind," a horse "kind," a dinosaur "kind," and so on.[9] Therefore, in less than 4,000 years post-flood, one generic marsupial pair and one generic monotreme pair made their way across a vast ocean to repopulate Australia, including diversifying into the many species living today. Creationists acknowledge the challenge but offer explanations.

Paul F. Taylor and others[10] propose a "little secret" as to how non-flying mammals spread: watercraft. Logs from massive forests of trees ripped up by raging floodwaters were floating about, creating little boats on which animals could hitch a ride to far-flung points on the planet. I've seen tongue-in-cheek quips about Noah "swinging by" Australia to drop off the kangaroos before heading back to Ararat or setting the kangaroos afloat in a dinghy, but apparently the creationist explanation isn't far off. In order to make the science fit Genesis, we have kangaroos in kayaks.

Other creationist explanations include post-flood land bridges. Michael Oard, a retired meteorologist who writes for multiple creationist institutions, argues for a massive post-flood ice age.[11] Using Oard's arguments, creationists extrapolate the existence of frozen land bridges, including to Australia.[12] Any evidence to the contrary is dismissed because "a biblical model of animal migration obviously must start with the Bible."[13]

What about the abundance of marsupials and dearth of placentals in Australia? Easy, according to Paul Taylor: marsupials were able to travel farther and faster than the placental mammals because they carry their young in a pouch. The pouchless placental mammals, on the other hand, had to stop constantly along the way to care for their young.[14] Consequently, the race to Australia was won by kangaroos in kayaks with their built-in baby wraps, while the placentals were left behind dragging along their tired and whiny toddlers. Taylor offers no explanation as to how the egg-laying monotremes beat the placentals to the Australian finish line.

Whether it was kayaks or land bridges, apparently the path to Australia was a one-way only, marsupials and monotremes only, no-placentals-allowed thoroughfare.

8

Written in Stone

William Smith was born into humble circumstances, orphaned as a boy, and then sent to live on his uncle's farm in Oxfordshire, England. His uncle was amused by his little nephew's love of rock collecting, especially the small clam-shaped rocks William and his buddies used as marbles. William also loved to collect "pound-stones,"[1] large rocks that looked like spikeless sea urchins and were often used to balance butter scales. No doubt about it, William Smith was a fossil enthusiast from boyhood.

Smith was a nineteenth-century working-class man with limited formal education. He worked as a surveyor's assistant and eventually as an engineer building roads and canals, always observing the rock layers exposed by excavations. Smith never lost his childhood fascination with fossils. Over and over again, Smith noticed that fossils were always found in sedimentary rock and certain fossils were always found in the same layer, or "strata" of rock. Amazingly, the order of fossils was always the same, no matter where he looked. From the bottom layers to the top layers, Williams could always predict which fossils he would find.

Wherever Smith examined rock layers, the succession of fossils was consistent. In 1815, Smith published a groundbreaking geological map of England and Wales, *Strata Identified by Organized Fossils,* followed by two more publications about fossils across England. Smith moved fossil collecting from a mere Victorian curiosity hobby to a real life, practical application in technology. Smith's "faunal succession"—specific fossils only occur in certain layers of rock—laid the foundation for modern oil and gas exploration.

William Smith and two fellow fossil-enthusiast friends, both clergymen, named over twenty-three rock strata along with the fossils they held. Smith was a man of faith, but he made no attempt to hide the facts of his discoveries or make them fit a biblical timeline. The facts were simply what they were. For Smith, fossils were insight into God's creative process: God must have created life in a manner that was progressively complex.[2]

Life's Photo Album

If evolution is true, if all life descended from a common ancestor, if "descent with modification" is fact, what kinds of things would we expect to see in the fossil record?

We would expect to see snapshots of the past that, when assembled chronologically, illustrate a panorama of evolutionary change over the past four billion years. This "photo album" in stone might be hazy in places and may have bits and pieces missing, but the pattern would be unmistakable. In reality, the fossil photo album we have is missing pictures here and there, primarily because fossilization is rare. Yet rare as it is, we have an amazing wealth of fossils.

When the brilliant (yet notably crabby) twentieth-century evolutionary biologist J. B. S. Haldane was asked what evidence would disprove evolution, he famously retorted “fossil rabbits in the Precambrian.”[3]

Haldane’s snarky response makes a valid point: the fossil record tells a story, and that story is predictable.

Precambrian rocks are the oldest rocks on earth. Precambrian life is overwhelmingly microscopic, mostly fossilized mats of simple, one-celled bacteria. In the topmost layers of the Precambrian, we find blobby multicellular organisms resembling sea pens and sea jellies. What is completely absent from the Precambrian is anything with a hard body or bones or legs or even a head. And most certainly, nothing with big ears and a cottontail.

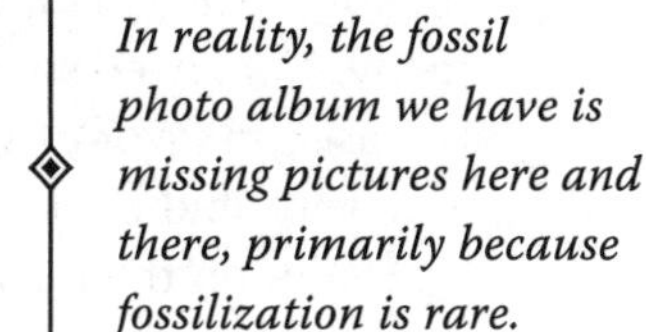

In reality, the fossil photo album we have is missing pictures here and there, primarily because fossilization is rare.

Most fossils form when an animal or plant dies and is covered with mud or silt. The sediment eventually hardens into rock, encasing the animal or plant inside. As the body deteriorates, minerals seep into the rock and replace the living tissues. Sometimes the hard parts completely deteriorate, and a cast of the body is left behind. Other types of fossilization preserve the entire animal, as happens when insects are trapped inside hardened resin.

We can also learn a lot from fossilized “left-behinds” of animals. Fossilized footprints and trackways tell us about the size and shape of an animal, the stance of the animal, and how it moved. Footprints and trackways tell us about behaviors, too. Did the animal travel alone or in groups? Did they travel with their young? Coprolites (fossilized feces) provide a wealth of information: What was the animal's diet? In what kind of ecosystem did the animal live? Coprolites can even tell us about the structure of the animal's digestive tract.

"Snapshots" from the early years of life are minimal. For the first 80 percent of life history, all living things had soft bodies. The fossil record is biased against soft bodies and biased in favor of species with hard parts, such as shells, bones, or teeth. Nevertheless, we have many imprints of soft-bodied animals and plants in rock-hardened mud. The fossil record is likewise missing many short-lived species and rare species. Fossilization is biased in favor of species that existed for a long time and were abundant and widespread.

Evolution theory says changes from one species to a different species, especially changes across classes or phyla or kingdoms, result from accumulated small changes over thousands, millions, and billions of years. Here is a common misunderstanding: evolution is a sudden event in which a fish gave birth to an amphibian, or a reptile gave birth to a mammal, or an ape had a human baby. Not only is this a misunderstanding of evolution, it is a misunderstanding of basic biology. If evolution is true, we would expect to see a continuum in the fossil record, not giant leaps.

Evolution theory says changes from one species to a different species, especially changes across classes or phyla or kingdoms, result from accumulated small changes over thousands, millions, and billions of years.

If all living things arose from an original ancestral cell, we would expect to see a progression from simple life forms to more complex. The panoramic story we see in the fossil record is a story of trends. We see a trend toward complexity in animals, from single-celled organisms in the oldest rocks to complex multicellular vertebrates in newer rocks. We see trends in body architecture, from animals with no heads, to animals with heads, to animals with heads and four limbs. We see trends in body configuration, from asymmetrical bodies (like sponges), to radially symmetrical bodies (like jellyfish), to bilaterally symmetrical bodies (everything else, from worms to humans). We see corresponding trends in plants also, from single-celled algae, to simple land plants, to plants with increasing complexity in structure and reproduction.

While we see *first appearances* of increasingly more and more complex body forms in the fossil record, the least complex do not disappear. There are trends in the fossil record, but primitive forms continue to live alongside more complex forms. All primitive animals and plants did not disappear when a more complex form evolved. We have plants and animals living today that are very much like their primitive ancestors, changing very little over time. Lampreys, coelacanths, and horseshoe crabs have changed very little since they are first identified in the fossil record. Ginkgo trees have changed very little in 100 million years. Once life forms appear, some of them continue throughout subsequent fossil layers, while others go extinct and vanish from the record.

Life: The Early Years

Just as William Smith discovered in England, there is a predictable succession in rock layers. And beyond Smith's England and across the planet, this observation holds true: older rocks are distinguishable from younger rocks by the fossils found within.

In many locations, rock layers lie neatly in chronological layers, like the layers of a cake. But in other places, movements in the earth's crust have disturbed the layers. Continental plates shift, mountains are pushed up in some places and eroded in others, rocks crack and slide, volcanic eruptions break the crust, and our neatly laid rock layers slide out of alignment. Looking at a broad view of a geographical area, geologists can recognize these shifts in alignment and piece the order back together. Additionally, we can determine the absolute age of rock (see Chapter 5) and then compare rocks of the same age from wide-ranging places on earth. Knowing both the physical succession and age succession in rock layers, we can construct a timeline of life.

The oldest fossils are mats of single-celled bacteria in rocks about 3.5 billion years old. For the next two billion years of earth history, bacterial mats are all we see. After that, we begin to find evidence of eukaryotes—more complex single-celled organisms with chromosomes contained within a cell nucleus. Up until this point, there were no multicellular organisms. In the oldest layers of rock, every living thing existed as a single cell.

In rocks around 630 million years old, we start to find multicellular organisms. Nothing with heads, much less eyes or limbs or organs, but primitive bodies nonetheless. These creatures left imprints of their soft bodies in hardened sediments. Known as the "Ediacarans," these animals were fern or feather-shaped (similar to modern sea pens, but not exactly), and some were wormlike.

And then—seemingly out of nowhere—around the 540-million-years-ago mark, we witness an "explosion" of life. Not just mats of single-celled organisms or the fortuitous imprint of soft-bodied Ediacarans, but a diverse display of animal life. Tiny little things with mineralized shells appear in abundance in the first rocks of this geologic era known as the Cambrian. Nicknamed the "little shellies," these tiny animals display a variety of shapes, textures, and spiky-ness.

After the "little shellies" layers, we find larger shelled animals and evidence of burrowing wormlike animals. The familiar insect/crab-looking trilobites with their easily fossilized hard body covering are plentiful in the next layers up. Finally, in the youngest Cambrian layers, we find teeming marine life: more varieties of trilobites, more varieties of shelled animals, primitive arthropods and echinoderms, wormlike creatures and other exquisite soft-bodied animals, diversified algae, and primitive skeletonless chordates. Chordates, by the way, eventually give rise to animals with bony skeletons (like us), but we are getting ahead of ourselves.

With such a "sudden" appearance of broad diversity, it is no surprise that this geological time period is often referred to as the "Cambrian explosion." However, *sudden* is a relative term—the Cambrian was a time of great diversity, played out over eighty million years. Donald Prothero calls the oft-cited "explosion" a myth and prefers the description "Cambrian slow fuse."[4] Despite the appearance of so many ancestors to modern life, the Cambrian includes nothing with true bones, limbs, or even heads.

Creationists have seized upon the "sudden" aspect of the Cambrian as evidence for the Genesis creation week. Some creationist writers acknowledge the Precambrian Ediacarans; others ignore or dismiss them. Even the "little shellies" of the early Cambrian are dismissed by best-selling creationist writer, Stephen Meyer.[5] Ignoring or downplaying Precambrian and/or early Cambrian organisms implies more suddenness, more "explosion" to the "Cambrian explosion." Regardless, all major creationist think-tanks agree: there are no precursor species in the Precambrian and diverse life appears "suddenly" in the Cambrian.

The creationist explanation for the Cambrian as well as for the rest of the fossil record is Noah's worldwide catastrophic flood. The same problems with a generalized sorting explanation apply to the Cambrian: no matter where in the world Cambrian layers are exposed, we find no mammals, no birds, no dinosaurs, no humans, not even any fish or amphibians. While diverse, Cambrian fossils are only headless and limbless primitive versions of marine creatures.

Traveling forward in time from the Cambrian, we find the first fishes. Initially they are few, primitive, jawless, finless, and covered with armored plates. But in subsequent layers, fish rule the day: fish with jaws and fins and skeletons made of cartilage, then fish with bony skeletons. We find fish with spiky rays supporting their fins and we also find fish with fleshy fins supported by rod-shaped bones. Both types have modern descendants, but ray-finned fish are most familiar to us. Some fish, changed little from their early ancestors, swim the oceans today: lampreys (jawless fish with a rudimentary backbone) and the sharks, skates, and rays (jawed fish with skeletons made of cartilage). While the fish ruled the seas, the first plants and insects ventured out onto land.

Still going forward in time, we find the first tetrapods, the four-limbed animals. Tied to water for egg-laying purposes, these primitive tetrapods nevertheless were exploring dry land at the water's edge. Before long, these early amphibians were the biggest game in town. Some were huge. Built like crocodiles, they were the apex predators of their day.

Other giants populated the forests: seagull-sized dragonflies and millipedes more than two meters long. And speaking of forests, lush growths of giant club mosses, horsetails, and ferns dominated. We also find the first seed plants, conifers bearing their seeds on cones. The world's rich coal seams, still fueling many parts of the world today, were produced by these swampy forests. In the oceans, fish swam with early echinoderms (like sea stars and sea urchins) and a variety of shelled mollusks, small and large.

The Game-changer

Onward. In rocks about 300 million years old, we find evidence of a game-changing trait in four-limbed animals. Like fish, four-limbed amphibians are tied to the water for reproduction. Eggs must be laid in water or they will dry out.

The key to a more permanent residency on land was the evolution of a closed egg: the amniotic egg. Amniotic eggs allow an embryo to develop tucked safely away in membranes and fluids, without the danger of drying out.

The first major group of vertebrates to lay their eggs on land were the reptiles. Able to move further away from the water, reptiles had a whole new world open to them. And wow, did they ever take off. After the first reptiles, we see early dinosaurs, more reptiles, then the more familiar dinosaurs, and the first birds. Some large predatory reptiles returned to an ocean life, their limbs modified as flippers to swim and hunt in the water. Other reptiles, the pterosaurs, took to the sky with skin-wings stretched over impossibly long fourth fingers.

In a lineage separated from the reptiles early on, we begin to find primitive mammal-like animals, still retaining some reptilian traits. During the heyday of the dinosaurs, we find the first true mammals, tiny and unimpressive, but mammals, nonetheless.

By about 100 million years ago, the myriad four-limbed creatures were crawling, running, and flying among the very first flowering and fruiting plants.

We have evidence of five major mass extinctions since life first arose on earth. The last mass extinction, the K-T extinction (also called the K-Pg extinction), was about sixty-five million years ago. Although it was not the largest mass extinction in geologic history, the K-T extinction decimated life on land and in the seas. After this time marker in the rock layers, we find no more (non-avian) dinosaurs, no more flying pterosaurs, and no more enormous reptilian ocean predators.

With the extinction of the domineering land and sea reptiles, the tiny and meek mammals took advantage of newly available living spaces. Mammals diversified all over the planet and many grew to enormous sizes. One group of mammals—the primates—first appeared around fifty-five million years ago. Initially small, the primates diversified into the new-world monkeys, the old-world monkeys, and the apes.

Around six million years ago and only in Africa, we find evidence of primates with some recognizable human traits. By three million years ago, we find primates walking fully erect, with humanlike hands and teeth. By two million years ago, we find primates classified in the genus *Homo*, the same genus as modern humans. The various *Homo* groups walked erect, had human hands and teeth, and used tools. About 1.8 million years ago, the first members of *Homo* left Africa. Around 200,000 years ago, we find the first fossils of modern humans in Africa. Modern humans left Africa about sixty thousand years ago and spread throughout the world, arriving in the Americas by fifteen thousand years ago.

Although this has been an abridged trip from the Precambrian to present day, the pattern in the fossil record is unmistakable: life in the oldest rocks is primitive and simple, followed by increased variety, diversity, and complexity in subsequent layers. Life has "tried out" a host of body forms; some survive into present time, others are extinct. And what's more, my simplified fossil journey did not include transitional species. In reality, the fossil record is full of "in-betweens"—mosaics with both primitive and modern features. On to the next chapter for transitionals!

9

In Search of the Missing Missing Link

Maybe it was in the spirit of "Keep Austin Weird," the motto adopted by the quirky and fun capital of Texas. On a bright September day in 2013, a Big Tex and a T-Rex trooped across the University of Texas campus, leading a small crowd. Marching behind a "Stand Up For Science" placard, the iconic Texas cowboy and the iconic Texas dinosaur made their way to the State Board of Education's hearing on the adoption of biology textbooks for public schools.

For decades, Texas has ruled textbook adoptions nationwide. Due to its size and large number of school districts, textbook publishers overwhelmingly seek to please Texas schools. The textbook at the center of the controversy that drew Big Tex and T-Rex and the other demonstrators was a widely adopted and highly regarded high school biology textbook. Six of the 28 state textbook reviewers rejected it due to multiple "errors" in the book—all related to the topic of evolution.[1]

After expert testimony addressing the "errors" one by one, the textbook was eventually adopted. Evolution remains, however, one of the most misunderstood and misrepresented science concepts.

Common Misconceptions

Misconception: Evolution is a theory about the origin of life. Scientists are interested in how life began, but the emergence of life is not a part of evolution theory. Once life arose, evolution explains how it spread and diversified.

Misconception: Evolution requires atheism. The theory of evolution says nothing at all about God, or religion, or any philosophy or world view.

Misconception: Evolution is not observable or testable. Evolution has been documented in controlled laboratory experiments as well as in studies of natural populations. There is a wealth of observable evidence. We do not have to be physically present when an event occurs in order to accept the reality of the event. Time and time again, evolution theory has been successful in predicting new discoveries.

Scientists are interested in how life began, but the emergence of life is not a part of evolution theory.

Misconception: Many scientists reject evolution. Evolution is accepted by the overwhelming majority of scientists—99 percent of active research scientists and working PhD scientists accept evolution.[2] To this date, there have been no peer-reviewed studies challenging the principles of evolution.

And then, of course, everyone's favorite: the elusive missing link.

Misconception: We have never found a missing link. Well, actually, this one is true. We have not found "a" missing link. In reality, there is not *a* missing link, but multiple thousands. Biologists call them "transitional species." Transitional species are not weird mishmashes like mermaids or a monkey-squirrel-fish but are mosaics with both primitive and advanced features. Transitionals document the evolution of species.

Just as a family photo album contains representative snapshots of a particular family member, the fossil record gives us snapshots in the life history of a species. We do not have every single species that ever lived represented in our photo album of life, and some steps or stages are smudged or missing, but we have more than enough to create an overwhelming picture of change over time.

Evolution of the Monkey-Squirrel-Fish

The common ancestry of all life is the most misunderstood and misrepresented aspect of evolution theory.

Although geology was the focus of the popular creationist film *Is Genesis History*,[3] common ancestry and transitional species also shared the stage. Featured speakers offered arguments ranging from the usual ("you can't build complexity one step at a time") to the absurd ("you aren't going to get a shark to evolve into a bird"). Common ancestry and transitional species were summarily dismissed by caricature:

"It's hard for me to imagine God being delighted seeing creatures flopping around on the ground, trying to produce wings, or trying to produce lungs."[4]

Is Genesis History repeats the creationist refrain of missing "missing links" and the non-existence of transitional forms in the fossil record. And, actually, there aren't any transitionals if you are looking for creatures flopping around trying to grow wings or lungs.

If descent from a common ancestor is fact, we would expect to find animals and plants in transition. We would expect to see animals and plants with primitive features and also more modern features. And furthermore, if common ancestry is fact, we could predict where such transitional organisms would be found.

From Water to Land

Dr. Neil Shubin of the University of Chicago is a fish paleontologist. Dr. Shubin's research passion is the transition from water to land. In 2000, Shubin and his team set out on the expedition of a lifetime, an expedition spanning more than six years. The goal of the team was to document the transition from fish to four-limbed land animal.[5]

If you were to choose a location on earth in which to spend four summers chipping away at rocks, where would you choose? Beach? Mountains? Anywhere with a pleasant climate?

Unfortunately, fossil-digging doesn't work that way. You can't just start digging willy-nilly in a fun vacation spot. First of all, you have to find the right type of rock because not all rocks bear fossils. Rocks formed by super-heating processes (lava, granites, marbles) are very unlikely to bear fossils, and certainly not fish fossils. Rocks bearing fossils are formed by sedimentation—limestones, sandstones, and shales. Second, the rocks have to be exposed to the surface. A wealth of sedimentary rock underneath your feet is of little benefit if the rock is buried under tons of soil or water.

Having the right type of rock exposed on the surface is a start, but for Dr. Shubin and his team, there was one more important criterion. Shubin knew the age of rocks where fish fossils appear. Shubin also knew the age of rock where the first four-limbed animals appear. If evolution is true, if common ancestry is fact, Shubin predicted that animals "in transition" between water and land would be found in rocks of an "in-between" age.

Using geological survey maps of the world, Shubin identified a location meeting all three criteria of type, exposure, and age. Over the next six years, Shubin and his team spent four summers on Ellesmere Island in the High Arctic of Canada. The Arctic Circle is not your typical summer destination, but if the fossil record tells a story, you must look where the missing piece of the story would be.

The breakthrough came when a team member cracked open a rock and discovered a snout staring back at him. As they removed more rock, they knew they were definitely looking at a fish, but this fish was not like any previously known fish. This fish had a flat head, the kind of flat head seen in crocodiles. And unlike any other known fish, this fish had a neck! This fish also had a shoulder, and inside its front fins were bones of an upper arm and a forearm, even a wrist. Shubin and his team named their find *Tiktaalik*, meaning "large freshwater fish" in the local Inuktitut language. Since the initial discovery, ten individual *Tiktaaliks* have been found.[6]

Tiktaalik was an incredible find, a wonderful example of an intermediate, or transitional animal. *Tiktaalik* was definitely a fish, closely related to lungfish living today. *Tiktaalik* had fins, gills, lungs, and scales. But *Tiktaalik* was a fish with traits of four-limbed, land-dwelling animals. *Tiktaalik* was a fish with a neck and shoulders, allowing movement of the head. *Tiktaalik* was a fish with sturdy ribs capable of supporting air-breathing and a pelvis more robust than we commonly find in fish. And remarkably, *Tiktaalik*'s front fin had the same bone structure we see in all four-limbed creatures, including humans: one bone, then two bones, then lots of little bones (including wrist bones), and primitive digit bones. *Tiktaalik* was a fish with a wrist and neck and could most likely prop itself up on its fins and walk in shallow water, but it is unlikely *Tiktaalik* walked on land.

Here is an important point to understand: *Tiktaalik* is a "transitional" or "intermediate" animal only from *our* perspective. In its day, it was the man of the hour. It was simply the animal it was. Its primitive tetrapod (four-limbed) traits gave it an advantage, so it survived. *Tiktaalik* could boost itself up above the shallow water surface using its front fins and look around with its neck. Only from our vantage point can we see *Tiktaalik* as a transitional species, a snapshot of the transition from water to land.

Here's a second important point to understand: *Tiktaalik* is not a "missing link" born one day from a regular everyday fish. *Tiktaalik* is a remarkable snapshot of the evolution of the tetrapod limb, but he's part of a continuum. In layers of rock just older than *Tiktaalik*, we find lungfish with flat heads and primitive forearm bones, maybe good for flopping a bit but not moveable enough to hoist up the body. Animals in rocks just younger than *Tiktaalik* are still quite fishy but have definite limbs. In these younger animals, not only do we find the limb bones we saw in *Tiktaalik* (one bone, two bones, wrist bones), but the limbs end in digits. Some had ears for hearing on land as well as in water.

Keep moving forward in time, and we find fish-ish tetrapods, but they no longer have gills or scales. The older the rock, the more fish-like and less amphibian-like the animal. The younger the rock, the more amphibian-like and less fish-like. *Tiktaalik* and his older and younger fishy cousins tell us that many of the features tetrapods use to live on land originally appeared in water-dwelling fish.

And on it goes. The earliest reptiles had many amphibian features. The earliest mammals were reptile-like. Interestingly, primitive features do not always disappear. My favorite example is the modern monotremes of Australia, mammals who still lay leathery eggs as did their reptilian ancestors.

It is tempting to imagine a simple linear path from ancient fish to modern four-limbed animals. In reality, there have been numerous branches splitting off from the others over evolutionary time, flourishing for millions of years, then going extinct. The modern four-limbed animals we see are descendants of ancient lineages that managed to survive. What's more, not all lineages of lungfish evolved into tetrapods. Modern lungfish are descendants of lineages that did not evolve tetrapod traits.

The strength of a science theory lies in its ability to predict. According to the theory of evolution, we would expect to see subtle transitions in the fossil record, not sudden leaps. Using the theory of evolution, we can predict where such transitionals will be found.

Sure enough, that is exactly what we see. When Neil Shubin went looking for a transition between water and land, he found it exactly where evolution theory predicted it would be found.

Feathered Flyers

For centuries, limestone quarried near Solnhofen in southern Germany was highly valued for its marble-like beauty. Romans used it for building and paving and it still adorns palaces across Europe. In the eighteenth century, lithography was popular as an emerging technology for mass producing illustrations, and lithographers prized the smooth, flawless surface of Solnhofen limestone for the process.

During the Jurassic period (200–150 million years ago), southern Germany was covered by a shallow sea, with networks of tropical and subtropical lagoons over the area of Solnhofen. A steady, quiet rain of microscopic plankton shells constantly fell on the basin floors, eventually hardening into beautiful limestone. As the tiny shells fell, they covered anything drifting in from the lagoons or the surrounding land. The water in the bottom of these quiet lagoons was still, salty, and stagnant—and thus very oxygen-poor. The remains of the animals buried in these oxygen-poor waters would not have been damaged by scavengers, and so they lay entombed in pristine condition.

As quarrymen checked slabs of limestone and set aside the most flawless pieces for lithography, they also set aside beautifully preserved fossils for collectors. In 1860, a perfectly preserved single feather was found, almost identical to the flight feathers of modern birds. Before long, multiple complete skeletons of the species to which this feather belonged were discovered. The fossils were beautiful imprints in stone with stunning details.

The species was named *Archaeopteryx lithographica,* a nod to the stone in which it was found. *Archaeopteryx* had wings, was covered in feathers, and had a furcula (the fused collarbones we call a wishbone). Bird, right?

But *Archaeopteryx* also had a full set of teeth, a long bony tail, claws at the end of its wings, dinosaur-type vertebrae, and a slashing claw on its hind feet. *Archaeopteryx* was a beautiful mosaic of dinosaur and modern bird traits. Originally hailed as "the first bird," *Archaeopteryx* is joined by a wealth of other feathered dinosaur discoveries. China has been a treasure trove of feathered dinosaur finds since the 1990s. Some are more dinosaur-like, some more birdlike, and some are in-betweens, like *Archaeopteryx.*

Archaeopteryx does not seem built for flight, although he may have achieved the kind of short-distanced flapping flight seen in partridges.[7] Feathers seem to be standard equipment for the theropod branch of dinosaurs, a group including not only *Archaeopteryx* and modern birds, but also velociraptors and *Tyrannosaurus rex.* Feathers are made of the same material as reptilian scales.

Originally, feathers were thought to be crucial to bird flight. However, feathers are found on theropod dinosaurs clearly not built for flight. Of what use are feathers to an animal that doesn't fly? Modern birds use feathers for flight, obviously, but feathers also provide a warm coat of insulation. Many modern birds are famous for their elaborate mating displays featuring vivid colors, iridescent sheen, and fancy feathered-fan dances. Feathers evolved before flight for warmth or mating displays or both.

Dinosaur to bird evolution is almost seamless in the fossil record. Dinosaur-bird evolution is so well-documented that modern classification systems now place birds within the larger reptile group, on the dinosaur branch of the reptiles. Specifically, birds are on the therapod dinosaur branch. Many modern biology books and texts refer to birds as "avian dinosaurs" and the traditional dinosaurs as "non-avian dinosaurs." Enjoy your Thanksgiving dinosaur, folks!

Return to the Sea

About fifty million years ago, a wolf-sized meat eater was prowling for meals along the shores of a shallow ocean in what is now Pakistan. Chemical analysis of its bones tells us fish was a dietary staple for this guy. Although his fossilized skeleton looks like other land-dwelling wolflike animals, oh what a big skull he had! It was quite elongated, with nostrils at the end of his long snout. Meet *Pakicetus*.

The biggest surprise, however, was hidden inside *Pakicetus*'s skull. A thickened dome of bone enclosed his middle ear. It wasn't the structure itself that was never-before-seen. There is a particular group of mammals having this bony ear structure, called an *involucrum*. Until *Pakicetus* was found, only this single group of mammals was known to have an involucrum.

Pakicetus, a four-legged, wolflike, land-living mammal had the unique ear structure of . . . a whale! *Only* whales (both living and extinct) belong to the exclusive "involucrum club."

Other "walking whales" have been found, most in and around Pakistan. A few million years younger, the crocodile-like *Ambulocetus* was larger and longer (about 11–12 feet) and had an elongated whale-shaped head. *Ambulocetus* had strong legs, flipper-like feet, and a large muscular tail, probably used in swimming. Bone analysis indicates he drank both fresh and saltwater, likely living by an estuary or in bays between freshwater and open ocean.

Over the next ten million years, whales adapted more and more to life in the water. Nostrils moved higher and higher up the snout, creating a blowhole for breathing. The trunk elongated tremendously, and tails acquired the fluke shape seen in modern whales. In some lineages of ancient whales, teeth are reduced, and we find the first evidence of filter-feeding seen in most modern whales.

By forty to thirty-three million years ago, hind legs were greatly reduced, tiny, and non-functional. One of my favorite things to see in natural history museums are the modern whale displays, especially if they have hanging skeletons. The hugest of the huge whales have teeny-tiny useless pelvises in their abdomens, not attached to anything. In museum displays, the useless pelvis is usually hung by thin wires below the whale, a reminder of the days when whales walked.

The whale fossil record is not a straight march from four-legged land mammals to modern whales. Modern whales (including dolphins) descend from a crew of mammals that walked on land, patrolled riverbanks, and swam in fresh water, estuaries, and oceans, each one adapted for his own place and time. But from our perspective, we can identify important transitional whales: four-legged land mammals, amphibious hunters, and finally, full-time ocean residents.

By the early 1990s, genetic testing revealed that modern whales are most closely related to artiodactyls, the even-toed hooved animals like camels, sheep, hippos, and antelopes. Amazingly, modern whales and dolphins are so closely related to hippos that whales, dolphins, and hippos share the same classification suborder. Molecular biologists continued to confirm the genetic connection throughout the 1990s, but to their great consternation, paleontologists were not having much luck confirming the artiodactyl connection from the fossil record.

As it happens, artiodactyls have a very distinctive bone in their ankles, shaped like a double pulley. All artiodactyls have this unique ankle—all of them—except of course the ankle-less modern whales and dolphins. And while we had many primitive walking whale fossils prior to 2000, none were found with complete limbs. Since 2000, however, multiple four-limbed fossil whales have been found with fully developed hind limbs, complete with double-pulley ankle bones. Using genetic evidence, evolution theory allowed paleontologists to correctly predict such an ankle would be found in primitive walking whales.[8]

A Very Bushy Family Tree

If you could identify your many-many-many greats of grandparents from a thousand years ago, and you also knew all of the lineages descended from them, you could create a family tree with all of your ancestors going back to that time. The tree would have very full branches, growing out in multiple directions. You could possibly trace a direct line from the many-greats grandparents to you, but most of your family tree is populated with indirect ancestors: aunts, uncles, and cousins.

A family tree from the distant past to you would spread out in many directions, with some branches full, other branches sparse, and a lot of dead ends. Many of your indirect ancestors died before they ever produced children and for others, all of their offspring died. Those branches of your family tree stop somewhere midway between then and now.

Evolution is not a straight-line, one-following-the-next march through time from an ancestor to a modern species. Was *Tiktaalik* the direct ancestor of the first four-limbed animal? Unlikely. Most likely, *Tiktaalik* was a cousin on the family tree leading to four-limbed animals. *Tiktaalik* tells us about the general trajectory of evolution from fish to four-limbed animals.

Transitional does not mean direct ancestry.

Indohyus and *Pakicetus* are both in the whale family tree, back when whales walked on land. Although *Indohyus* had the ear bones unique to whales, he was a small plant eater. *Pakicetus* was larger and more whale-like. But here's a surprise: the more primitive *Indohyus* is *younger* than the more whale-like *Pakicetus*. Nothing about evolution prevents a more primitive form of an animal living after or alongside an animal with more modern features. Both *Indohyus* and *Pakicetus* are transitionals, but it is unlikely that either are direct ancestors of modern whales. *Indohyus* and *Pakicetus* are cousins on the whale family tree, and both tell us about the general trajectory from land to sea.

Within specific lineages, we see fits and starts and successes and failures, with winners and losers often overlapping each other. Species with modern traits coexist with ancestral types.

The modern horse is an impressive example of "overlapping" evolution. Horse evolution is one of the best documented snapshots of the past we have. Most notable are changes in body size (from tiny to large) and foot structure (from three and four toes to a single-toed hoof). Although we see a trend over the last fifty million years from tiny multi-toed scamperers to large single-hooved gallopers, various lineages of horses overlapped each other. Branches of the horse family tree (tiny, medium, and large) lived and flourished for millions of years, then went extinct. Today, only *Equus*, the genus of the modern horse (including zebras), remains.

If evolution is *not* true, if all living things were specially created in one week's time, what would we expect to see in the fossil record?

Is that what we see?

Conducting the Orchestra

It's opening night of a new symphony series, and there's been a last-minute change in the program. You have a choice: you can whisper the program change into the ear of the nearest trombone player, or you can whisper the change to the conductor. Who do you choose? The conductor, of course. If you want to effect a dramatic change in a single move, you go to the one in charge of the action.

The (apparently) sudden and dramatic appearances of many new animal forms in the Cambrian fossil layer perplexed Charles Darwin. Darwin was at a disadvantage, living as he did prior to the discovery of fossils older than the Cambrian and prior to the age of modern genetics.

The evolution of novel structures and new body plans still perplex many of us. Christian apologist and author

Frank Turek, in a 2020 interview with Capturing Christianity, makes this bold statement: "you can't take a genetic code and modify it and get new body plans."[9]

The "sudden" diversification seen in the Cambrian is often taken as evidence of special creation. The argument goes like this: Cambrian creatures, individually designed and created by God during the six-day creation week, have no ancestors in the fossil record. Precambrian fossils exist, but are usually ignored or downplayed in creationist discussions, further confirming the "suddenness" of the Cambrian. Other novel structures or new body types arising after the Cambrian (such as whale bodies and limbless snakes) are likewise the result of special creation. According to the creationist/intelligent design model, randomly occurring mutations to a single gene could not produce the dramatic changes we see in organs, tissues, and body forms.

If the nineteenth century was the century of engineering and chemistry and the twentieth century was the century of physics, the twenty-first century is powering forward as the century of biology. Nowhere is biology's dynasty more apparent than in modern genetics. We've known since 1952 that DNA carries the instructions for building life. With the advent of gene mapping, our knowledge is exploding, with applications following close behind.

Here is one surprising discovery: the amount of DNA in an organism does not correlate with complexity. Humans have about the same number of genes as do simple flatworms and flies. Rice and corn have more than twice the number of genes as humans. The vast majority of genes do not code for proteins, the building blocks of our bodies. In humans, less than 2 percent of our DNA codes for proteins.[10] What, then, is going on with the other 98 percent?

The 98 percent are conducting the orchestra.

What we originally (and mistakenly) called "junk DNA," is anything but. These genes aren't coding for the protein building blocks of bodies, they are controlling how bodies are put together.

One group of control genes found in animals as diverse as flies, mice, and humans orchestrates head-to-tail organization.[11] "Put the head on this end! Put the tail on that end! Put the body in the middle!" Another group of genes controls the number and placement of limbs. Other genes control timing; they are responsible for switching genes "on" or "off" at different points in development.

The tremendous increase in diversity during the Cambrian and beyond is no longer such a mystery. Turn off a control gene during leg development, and a crustacean ancestor with many legs gives rise to insects with six legs. Change the timing in limb development, and a land-dwelling, four-limbed ancestor gives rise to descendants with increasingly paddle-like front limbs and eventual loss of hind limbs.

A mutation in a single protein-building gene may be helpful or harmful or neutral. A mutation in a powerful control gene may also be neutral, but if not, the results can be dramatic. A control gene, turned on or off at the wrong time or turned on in the wrong place, has the power to create new structures or new functions for existing structures.

And could it get any cooler? We know versions of these control genes are found in all animals. Here's an example: the genes controlling limb development in snakes are turned off (no surprise there). Researchers replaced the limb-control genes in a mouse with snake limb-control genes. The resulting offspring? Mice with a head, a tail, and no legs.[12]

The evolution of dissimilar species from a common ancestor is hard for many to accept. The more we know about control genes, the more we understand how it happened.

10

It's All or Nothing: Intelligent Design

Caked with mud and riddled with holes, the only thing missing was the smell of sweat from a long day's labor. These jeans are the jeans of a working man, a man who is daily out in the trenches and up to his waist in dirt and muck as he earns his living by the sweat of his brow. These jeans testify to hard work by the wearer, dirty and physical.

Although every detail of these jeans indicates a long history of hard and dirty manual labor, these jeans are, in fact, brand-new. Anyone with an extra $425.00 could buy a pair.

These brand-new jeans, an offering from Nordstrom's department store, actually have no work history—they have simply been designed to look as if they do. Nordstrom's was soundly ridiculed for their tone-deaf "fashion statement," but no commentator was more scathing than Dirty Job's Mike Rowe:

"Finally—a pair of jeans that look like they have been worn by someone with a dirty job . . . made for people who don't."[1]

A Price Too High

For many people of faith, traditional young earth creationism demands too high an intellectual price, primarily because it demands a 6,000 to 10,000-years-old universe. A young universe and a young earth require the dismissal of modern geology, modern physics, and the sequential nature of the fossil record, and for many, that is a bridge too far. "Creationism" and "creation science" have been jettisoned in many religious circles and replaced with the intelligent design model. Intelligent design (ID) is infused with scientific vocabulary and complex concepts and has no objection to evidence from physics or geology. Intelligent design accepts the evidence for the age of the earth and the fossil record as a historical sequence of life.

Evolution itself is the problem. According to intelligent design, an unguided natural process could never produce the intricacies and complexities we see in living things. Only an "intelligent designer" could produce such awesome complexity. The unnamed designer specially designed (created) each organism in every detail and aspect. Although the designer is never identified in most intelligent design publications, without a doubt, the mysterious designer is the God of the Bible.

Any evidence indicating common ancestry is dismissed as part of the design, purposely added by the designer, like the pricey jeans that look like they have a history when in fact, they do not. Most intelligent design proponents concede *microevolution* (small changes in species, allowing for adaptions like beak shape in birds) but reject *macroevolution* (descent of species from common ancestors).

A Brief History of the Intelligent Design Movement

In a 1987 ruling (*Edwards v. Aguillard*), the Supreme Court of the United States ruled that creation science is a religious doctrine and therefore has no place in the science curriculums of public schools. Creationism in the public classroom was down, but it wasn't out.

Creationists regrouped, and out of the regrouping was born the intelligent design movement. The godfather of the movement is retired Berkeley law professor Phillip Johnson. Despite having no training in sciences, Johnson turned a questioning eye toward evolution. He offered no evidence or theory or proof of his case, just problems (as he saw them) with evolution. Johnson's approach was an attorney's cross-examination. He did not attempt to prove his case, he only needed to sow doubt. His 1991 book, *Darwin on Trial*, rekindled the fervor of anti-evolutionists. Johnson declared: "Science knows no natural mechanism capable of accomplishing the enormous changes in form and function required to complete the Darwinist scenario."[2]

Biochemist Michael Behe picked up the mantle of intelligent design leadership with his 1996 book, *Darwin's Black Box*.[3] Behe's dispute with evolution resides primarily in the fine details of life: the cell and biochemical functioning.

According to Behe, biochemical systems and cell structures are so complex, no natural process could have produced them. Take away just one step in a biochemical pathway or one small piece of a cellular machine, and the whole system collapses. Like a biological Rube-Goldberg machine, each piece and part must be present all at once in order for molecular and cellular processes to work.[4] Any intermediate form of a system would be nonfunctional. Intelligent design is all or nothing. Behe coined the term *irreducible complexity* to describe his "all or nothing" concept. Irreducible complexity became (and remains) the centerpiece of the intelligent design movement.

In 2005, another landmark religion-in-the-science-classroom case (*Kitzmiller v. Dover*) was heard in a United States Federal Court in Pennsylvania.[5] This time, it was intelligent design on trial. The school board of Dover, Pennsylvania, required intelligent design be taught alongside evolution in science classrooms. Witnesses for the board argued that intelligent design is science, and as such could legally be taught in public schools. The Dover school board planned to adopt *Of Pandas and People*, an intelligent design textbook for high school biology classes.

The goal for the pro-ID side was to present intelligent design as legitimate science with no religious underpinnings. Michael Behe was the key witness representing intelligent design.

Behe's testimony, of course, focused on irreducible complexity, and specifically on the immune system. According to Behe, the immune system is so complex, missing just one tiny component would render the entire system useless. Behe was adamant: an evolution explanation for the immune system would never be found. Behe was then presented with fifty-eight peer-reviewed publications and textbook chapters, all explaining the evolution of the immune system. His response? "It's not enough."[6] Behe eventually conceded the point: there are no pertinent peer-reviewed studies demonstrating design in any biological system.[7]

The final blow in the trial came from the proposed textbook, *Of Pandas and People*. The publishers were subpoenaed to provide early drafts of the text. Without a doubt, the original draft was written specifically to teach creationism.

Following the 1987 ruling against creationism in the classroom, the *Pandas* publishers produced a new draft of the text. A key witness in the Dover trial superimposed passages from the original drafts with the same passages from the revised draft. Wherever the terms "creation" or "created" or "creationism" were found in the original draft, the new draft substituted a form of the word "design." All other verbiage remained the same.[8]

The presiding judge was John E. Jones III, a conservative Republican appointed to the federal bench by George W. Bush. Judge Jones ruled that intelligent design does not meet the criteria of legitimate science and is inherently religious. As such, teaching intelligent design in public schools is a violation of the First Amendment of the U.S. Constitution.[9]

Although the overt teaching of intelligent design in public schools has been settled in the courts (for now), the struggle continues in more subtle ways, usually at the school board level in states and local districts. At the heart of these curriculum battles is inclusion of phrases like "teach the controversy" or "teach the weaknesses" of science theories. No one is advocating to teach the weaknesses or controversies of germ theory or gravity theory or atomic theory. The unspoken but blatantly obvious theory in question is evolution theory.

When evolution is depicted (as it often is) as godless, cold, uncaring, and random, the idea of design by an intelligent Designer is attractive. The contrast is not by accident. Early on in the intelligent design movement, Phillip Johnson showed his hand: "The objective is to convince people that Darwinism is inherently atheistic, thus shifting the debate from creationism vs. evolution to the existence of God vs. the non-existence of God."[10]

Designed Antiquity

Intelligent design accepts the age and chronological order of the fossil record. Although the sequence is accepted, organisms appear in the fossil record abruptly and likewise disappear and are replaced abruptly, with little to no evidence of "transitional" forms.[11] In other words, the designer specifically and separately crafted each iteration of every species that ever lived on earth.

Consider the evolution of the two living elephant species, the Asian elephant and the African elephant. Over the last fifty million years, more than 300 species of proboscideans (trunked mammals) have lived on our planet, some as small as pigs and rabbits. It was not a straight march to modern elephants; instead, we find a many-branched elephant family tree with different body sizes and trunk structures. Elephant branches often overlapped each other in time; twenty-two distinct species lived in just the last six million years.[12]

Likewise, modern horses (including the zebras) have a multi-branched ancestral tree. Over the last fifty-three million years, horses have been tiny, large, and intermediate in size. Some were adapted for forest living, others for woodlands, still others for savannahs or prairies.

Intelligent design demands separate and independent designs for each elephant and each horse in their respective family trees. Intelligent design demands each elephant and each horse appear abruptly on the planet, without a relationship to an ancestor species. Intelligent design demands that the appearance of actual relationships between extinct species and living species is just that—an appearance. Apparent shared ancestry is designed, a creative choice on the part of the Designer. The facts of natural history, Kenneth Miller observes, require a designer "who creates successive forms that mimic evolution."[13]

In fact, the fossil record boasts a wealth of ancestral histories for multitudes of species. We see beginnings of tetrapod (four-limbed) animals in lobe-finned fish, and we see the same limb arrangement repeated in each and every tetrapod, living or extinct. Regardless of the tetrapod—bats, cats, birds, whales, dinosaurs, frogs, pterodactyls, humans—we see a version of the tetrapod limb pattern: one bone, two bones, lots of little bones, digits.

Sometimes limbs or digits are reduced or elongated, sometimes fused, and sometimes lost, but the same tetrapod limb architecture is found in each and every tetrapod, nonetheless.

Intelligent design says the common limb pattern seen in all tetrapods, living or extinct, is simply coincidental—a design choice. The limb pattern is not inherited from a common four-limbed ancestor. In other words, the hind limb buds that grow (then reabsorb) in embryonic whales are a designed feature. Whales have flippers, not hands, but inside the flipper, you'll see five finger bones—again, a design choice.

Intelligent design explains these apparent relationships as the mark of a common designer who employs similar designs in similar ways. As a result, species appear to have a history that they in fact they do not have—just like our (brand-new) muddy and worn-out designer jeans.

Design and DNA

Below the surface—all the way down to the DNA level—there are a host of troubling issues for intelligent design.

Vitamin C is essential for life. Fortunately, most living things are able to make their own vitamin C. Monkeys, apes, and humans, however, cannot make vitamin C and must obtain it from their diets. Without vitamin C, humans develop a painful, debilitating, and often fatal disease called scurvy. Before the nineteenth century, sailors on long voyages were particularly susceptible to scurvy until it was discovered a daily dose of lime juice provided the necessary vitamin C.

An essential gene called the *GLO* gene makes vitamin C production possible. Humans, monkeys, and apes actually have the *GLO* gene in their DNA, yet they cannot make vitamin C. As it turns out, the *GLO* gene in humans and in their close primate cousins is broken; the gene is completely nonfunctional. Moreover, the *GLO* genes of humans, monkeys, and apes are broken in the *same* way.

Evolution easily explains the mystery: in an ancestor on the primate branch of the mammal family tree, a mutation occurred in *GLO*, rendering it nonoperational. From that point forward, all descendants inherited the nonoperational version of the gene. According to intelligent design, the Designer purposefully included an identically broken gene in the otherwise functioning DNA of humans, monkeys, and apes.

The case of the *GLO* gene is not unique. In egg-laying animals, there are three genes responsible for making the yolk needed to nourish the developing embryo before hatching. Mammal DNA has all three yolk-producing genes, but all three genes are broken. Again, evolution provides the answer: in a mammalian ancestor, the yolk genes mutated, rendering them nonoperational. All descendants from that point inherited the broken versions of the genes. According to intelligent design, the Designer intentionally inserted broken versions of yolk-making genes into the DNA of non-egg-laying animals.

Intelligent design explains the thousands of broken and nonfunctioning genes littering the genomes of humans and all other organisms as purposefully inserted by the Designer into otherwise functioning DNA. Any apparent inheritance of a gene (functional or otherwise) from a common ancestor is design mimicking evolution.

Michael Behe departs (somewhat) at this point from the conventional intelligent design model. To the dismay of many creationists (for whom he is a hero), Behe accepts common ancestry of all life for "the purposes of argument"; however, he does not believe common ancestry "explains anything."[14]

Irreducible Complexity: Nature's Rube Goldberg Machine

While Phillip Johnson's interpretation of intelligent design relies on lawyerly argumentation ("design requires a Designer"), Michael Behe's version is far more rooted in scientific terminology. Behe's defense of intelligent design relies on the intricacies of biochemical pathways and complex structures in living things. According to Behe, all parts and pieces and steps and structures of systems must be in place all at once, or the system is nonfunctional. Likewise, component parts and pieces have no function apart from the complete system. Just like a Rube Goldberg machine, take away just one piece and it all falls apart. Therefore, Behe argues, stepwise evolution by natural selection could never produce the complexities of life.

Intelligent design's go-to analogy for irreducible complexity is the common household mousetrap. A mousetrap is a simple machine consisting of about four component parts. Take away any one of the parts (the spring, for example), and the mousetrap is useless. All component parts must be present in order for the trap to function according to its designed purpose—catching a mouse.

Cell biologist Kenneth Miller (Brown University) has been the point man countering intelligent design since before the 2005 *Kitzmiller v. Dover* court case. Miller is the coauthor of a bestselling biology textbook and the author of two books specifically addressing intelligent design. Miller is also a practicing Christian.

Miller, in a true story from his childhood, addresses the mousetrap analogy.[15] Using a broken mousetrap with several parts missing, Miller's classmate built a functioning spitball catapult, capable of launching a juicy one from the gym floor to unsuspecting students in the balcony. Although missing key components, the "reduced" mousetrap was not useless. The reduced mousetrap caught no mousies, but it was still functional: it was an ideal spitball launcher.

The concept of irreducible complexity fails when it comes to mousetraps, but a mousetrap is not a living, biological system.

A Royal Disease

The British Queen Victoria was mother to nine children, who all amazingly (by nineteenth-century standards) lived to adulthood. Victoria's five daughters married into royal houses across Europe, and since royals only married royals, these marriages were often to fairly close cousins. Unknown to Victoria, she carried a deadly gene on one of her X chromosomes—a mutated gene in the blood-clotting process. Missing this vital step in proper blood clotting can be disastrous. Fortunately for Victoria, she was female and therefore had a second X chromosome carrying the functional version of the blood-clotting gene. Males, having only one X chromosome, aren't so lucky if they inherit the broken gene.

Though unaffected themselves, Victoria's daughters, granddaughters, and great-granddaughters passed the broken gene throughout royal families across Europe. Many of Victoria's male descendants, including one of her sons and three of her grandsons, bled to death in childhood or early adulthood. This form of hemophilia is an "X-linked" disease: females carry it, but mostly males are afflicted.

The vertebrate blood-clotting mechanism is a frequent example of irreducible complexity. To be sure, blood clotting in vertebrates is undoubtedly complex.

It consists of multiple steps, each dependent upon previous steps. If one step is missing or defective, normal blood clotting does not occur—thus the misfortune of Victoria's male descendants. A system requiring all parts at once in order to function could not have evolved, right? According to intelligent design, a partial pathway would be useless.

The evolution of the vertebrate blood-clotting cascade has been extensively researched since the 1960s. In whales and dolphins, we discovered a missing clotting factor, previously thought to be present in all vertebrates. We discovered three missing factors in puffer fish. Primitive, jawless lampreys lack multiple components of the vertebrate blood-clotting system. The conclusion? "Partial" pathways exist in modern vertebrates, and they function just fine.[16]

Let's go back a step and look at the chordates, the animals from which vertebrates arose. Sea squirts, a group of modern chordates, do not have any functioning clotting factors. What sea squirts *do* have are all of the protein building blocks from which those factors are built.

Now, go back one more step. The animals most closely related to chordates are the echinoderms: sea stars, sea cucumbers, sand dollars, and sea urchins. Echinoderms produce a protein clearly related to a key vertebrate clotting protein called fibrinogen. In echinoderms, however, the protein is not used to clot blood; echinoderms do not even *have* blood.[17] Yet, there it is, a precursor to fibrinogen, doing some other job in starfish.

Intelligent design says natural processes cannot produce the complexity of the vertebrate blood-clotting mechanism. On the other hand, if evolution is true, if the complex vertebrate blood-clotting system evolved, what kinds of things would we expect to see?

We would expect to find some vertebrates (like whales, dolphins, and puffer fish) missing parts of the system. We would expect to find primitive vertebrates (like lampreys), with related, but simpler clotting systems. In animal lineages closely related to vertebrates (like sea squirts and echinoderms), we would expect to find the raw materials for building clotting factors before we find actual clotting systems.

And that is exactly what we find.

Reduced Mousetraps

The showpiece and "Exhibit A" of irreducible complexity (all or nothing) has for years been a structure in the humble one-celled bacteria: the bacterial flagella. Bacterial flagella are like little outboard motors, propelling bacteria through their environment. And indeed, the bacterial flagellum is an amazing and complex structure. Flagella-powered bacteria can swim hundreds of body lengths in a second and they can turn on a microscopic dime. But are bacterial flagella all-or-nothing cellular machines? Must all constituent parts be present, and all at once, in order to be useful?

Flagella are not the only impressive sub-cellular machines in bacteria. Disease-causing bacteria have a piercing needle-like structure they use to punch holes in the cells they infect. Turns out, the proteins used to assemble the bacterial needle are identical to the proteins used to assemble the "propeller" part of the flagellar motor. The genes coding for the common proteins clearly indicate the two types of bacteria share common ancestry.

Intelligent design says that each subunit of the bacterial flagellum has no function apart from the whole motor. Yet, here is a subunit, not functioning as part of a motor, but functioning quite well as a toxin-delivering, cell-puncturing needle in related bacteria. While we do not have the entire flagellar puzzle solved, we do know this: it is not all-or-nothing. Bacterial flagella are not irreducibly complex.

On the other hand, if evolution is true, what kinds of things would we expect to see? We would expect to find protein machinery used for a purpose in an organism, then find the same machinery co-opted and repurposed for another function in a related organism.

And that is exactly what we find.

Evolution is a tinkerer. Nature is thrifty. Evolution does not start from scratch. Structures and processes are not built (or designed) from the ground up. Any modification in a species arises from genes already there, inherited from ancestors.

And on it goes.

Choanoflagellates are simple, one-celled aquatic protists. They produce a protein once thought only produced by multicellular animals. The protein (called cadherin) "glues" cells to other cells, making multicellular bodies possible. What is a simple one-celled organism doing with such a protein? As it happens, choanoflagellates do not use cadherin to stick to other cells; instead, they use it to snag food particles floating by.

Evolution is a tinkerer.

Before multicellular bodies existed, the genes existed to make cadherin. Evolution is a tinkerer. Nature is thrifty. Evolution does not start from scratch.

There's more.

The protein crystalline makes up the lens of the complex vertebrate eye. Crystallin is also made by a vertebrate ancestor, a boneless, brainless, headless, marine animal called a sea squirt. Sea squirts use crystalline to sense gravity. We don't know why the mutation occurred that resulted in early vertebrates producing crystalline in their eyes, but we know they already had the genes to do so.

The genes for building flowers are older than flowering plants.

The genes for building wings and feathers are older than bird flight.

Evolution is a tinkerer. Nature is thrifty. Evolution does not start from scratch. If evolution is true, we would *not* expect to find built-from-scratch pathways and structures.

We would expect to find partial pathways in earlier lineages leading to complex pathways in later lineages. We would expect to find parts, pieces, and raw materials with functions in earlier lineages different from their functions in modern lineages, but functioning, nonetheless. We would expect to find jerry-rigged genes and makeshift arrangements.

And that is exactly what we find.

11

You Can't Make a Monkey Out of Me: The Touchy Topic of Human Evolution

Clearly, the scientists in my family go WAY back. A newspaper ad (circa late 1800s) features my great-great-grandfather, Stephen Kellogg.

The self-bestowed credential of "professor" apparently qualified Professor Kellogg as a "scientific masseur" and "suggestive therapeutist." Family lore says his wife subsequently left him, not keen on the idea of her husband seeing the local townswomen in various stages of massage-necessitated undress (not to mention the wide possibilities of suggestions in "suggestive therapy").

The professor is an interesting node on my family tree. And branching off all around him are greats and greats of grandparents, aunts, uncles, and many-times-removed cousins. My family tree tells me that I am descended from the illustrious Professor Kellogg—he is my direct ancestor and I am his direct descendant. All the aunts, uncles, and cousins are my relatives, some more closely related than others. They are all my relatives, but I am not the direct descendent of all.

Our Common Ancestor

Dayton, Tennessee, 1925. Preachers with Bibles held aloft are evangelizing on street corners, vendors are hawking souvenirs, little girls are carrying monkey dolls, and a chimpanzee named Joe Mendi is sipping a Coca-Cola in the drugstore wearing a plaid suit, a brown fedora, and white spats.

John Scopes, a Tennessee high school teacher, was tried and convicted in the most famous science-versus-faith case to date: The Scopes Monkey Trial. Conventional wisdom says Scopes was in trouble for teaching evolution, but in fact, he wasn't in hot water for teaching that plants or even animals evolved over time. Scopes was found guilty of violating Tennessee's Butler Act, which forbade teaching against the "Divine creation of man" or that man "has descended from a lower order of animals."[1]

Teaching evolution per se didn't get John Scopes in trouble but bringing in the monkeys did. Many twenty-first century Americans feel the same way. More Americans are willing to accept evolution if it applies only to plants and animals than if humans are added into the mix.[2]

Here's the million-dollar question: "If we came from monkeys, why do we still have monkeys?"

A Twitter feed I follow is called "Take That Darwin." It trolls the twitterverse daily and retweets all the variations of the "why are there still monkeys" meme along with snarky responses ("Wow! Have scientists never thought of that??"). Irritainment, I know.[3] Here's the short answer—people did not "come from" monkeys.

Monkeys are still around because monkeys did not "change into" humans. Monkeys are not evolving into humans because they are doing just fine as monkeys, thank you very much.

My cousins and I share a common ancestor—our great-great-grandfather, "Professor" Kellogg. No one has ever asked me this question: "If you and your cousins all came from your great-great-grandfather, why do you still have cousins?"

When I was born, my cousins did not disappear. My cousins went about their lives, growing their own family trees.

Humans share a common ancestor with the great apes, most closely with chimpanzees. Closer human "cousins" have lived in the past, but all are now extinct. The chimpanzee is our closest living evolutionary cousin. Genetic analysis estimates the last common ancestor of humans and chimpanzees lived between eight million and five million years ago. After that, the two family trees branched off in different directions. Modern chimpanzees descended from their (now extinct) direct ancestors as well as adjacent branches of great aunts and uncles and cousins; modern humans descended from their (now extinct) direct ancestors as well as adjacent branches of great-aunts and uncles and cousins.

Family Tree or Family Bush?

Until recently, the record of human history was fairly straightforward. The human family tree (*Homo*) was scraggly—basically just a trunk and one or two branches. A few million years ago, some primitive branches of the family tree left Africa. These early migrants thrived for hundreds of thousands of years in Asia and Europe until a new species of *Homo* charged out of Africa and took the planet by storm.

The new kids on the block were us—modern humans—*Homo sapiens*. We were so good and smart and talented and verbal we out-competed or killed off all other *Homo* species until we were the last group standing, approximately thirty thousand years ago.

Or so we thought.

Turns out, the human family tree is a bit bushier—not quite the straight shot we once thought from chimpanzee to *Homo erectus* to Neanderthals to us. This conception was not helped by the famous *March of Progress* illustration, originally published by Time-Life in 1965.[4] You know the one: a single file line-up of fifteen, from a tiny monkey at the tail end of the line to a tall, handsome, perfectly postured modern human at the front end.

Turns out, the human family tree is a bit bushier—not quite the straight shot we once thought from chimpanzee to Homo erectus *to Neanderthals to us.*

This illustration has been reprinted and repeated, respected as science fact and disparaged as the march to godless evolution. Both interpretations are wrong. In reality, the *March of Progress* is no more scientifically correct than the Homer Simpson version, starting with the Homer-faced "Monkius eatalotis" and ending with "Homersapien." The evolution of modern humans was not an all-in-line march to the finish as the famous illustration implies. Modern humans sit on one tip of a branch of an ancient human family tree, a tangled tree with many branches.

All of the other branches in this tangled tree have died out. We alone survive.

But in the not-so-far past (relatively speaking), this was not the case. In the past, modern humans shared the planet with some of the now extinct branches of our family tree.

Enter the Hominins

Here's the real story: about 4.4 million years ago, the very first hominins (human ancestors) arose in east Africa. Although not the *first* early hominin, the *most famous* early hominin is "Lucy"—her scientific name is *Australopithecus afarensis*.

Lucy and the other *A. afarensis* were small—between three and a half to five feet tall. Their skulls were small and chimp-like, housing a small brain. Their teeth, however, were more human—small canines with arched tooth rows.

The bodies of Lucy and her kin were mosaics of ape and human traits. They had the long arms and curved fingers found in chimps who climb trees and chimp-like ribs. But their spine, pelvis, legs and feet indicate *A. afarensis* walked upright on two feet. In an amazing find in Laetoli, Tanzania, we have a trackway of about seventy preserved footprints made by three *A. afarensis* individuals. These footprints tell us Lucy's kin walked upright in a stride that was far more human than ape-like.

Our Closest Human Cousins

About 2.2 million years ago, our genus, *Homo*, arose in east and southern Africa.

Early *Homo* groups were less and less ape-like and more and more humanlike, compared to Lucy and the early hominins before her. Members of *Homo* were the first to have humanlike body proportions. Compared to their torsos, members of *Homo* had shorter arms and longer legs. Early members of *Homo* walked and ran like modern humans. Early *Homo* groups had larger faces, smaller brains, and more primitive teeth compared to modern humans, but *Homo* groups used tools, hunted, butchered meat, controlled fire, and probably cooked.

About one million years ago, some *Homo* groups left Africa and moved into Asia. Separated from their kin in Africa, new *Homo* groups arose and spread into Europe and across Asia, as far as Indonesia and Siberia. Compared to older *Homo* groups, the younger *Homo* groups had larger brains and fewer primitive characteristics.

Out of Africa, the ancestral *Homo* groups further split into at least two important groups: the Neanderthals and the Denisovans, and a possible third group, *Homo floresiensis*. These (relatively) recent branches of the human tree are very closely related to us and not all that different. These branches not only intersected over time with us, but in some cases, they shared physical space.

The most mysterious of the recent *Homo* branches is *Homo floresiensis*, dating to 190,000 to 50,000 years ago and found only on the island of Flores in Indonesia. *H. floresiensis* was tiny—only about three and a half feet tall. They made tools and hunted and may have used fire. Were they tiny when they arrived on the island? Or is their diminutive size due to long-term isolation on an island? We aren't sure.

The Neanderthals, the best-known human cousins, were in Europe from 200,000 years ago to as recent as thirty-five thousand years ago. They ranged from Wales in the west to as far east as Siberia. Neanderthals were shorter and stockier than modern humans, with bone structure and muscles adapted to cold environments. Their brains were as big as ours, often bigger. They used tools and fire, they hunted, lived in shelters, wore clothing, made jewelry, and sometimes buried their dead with offerings. Their ability to use language appears to have been limited.

The Denisovans are the most recently discovered close human cousins. In 2008, paleontologists digging in the Denisova cave in Siberia found human remains that were not genetically Neanderthal, nor were they modern human. The remains were a separate *Homo* group, but closely related to both Neanderthals and modern humans. Since the initial discoveries in the Siberian cave, more Denisovans have been found in Tibet.

The Denisova cave yielded an astonishing new find, first announced in August 2018. A bone from a 13-year-old girl (nicknamed "Denny") found in the cave stunned scientists.[5]

Denny had *equal* amounts of Denisovan and Neanderthal DNA.

Humans have twenty-three unique chromosomes. But we have two copies of each chromosome, one from biological mom and one from biological dad. In each of Denny's chromosome pairs, one chromosome came from an exclusively Neanderthal parent and one came from an exclusively Denisovan parent. Additionally, humans have a tiny bit of DNA in the mitochondria of their cells. *All* mitochondrial DNA comes from mom. Denny's mitochondrial DNA is Neanderthal.

It was as if we had a front-row seat. Denny was a first-generation offspring of a Neanderthal mom and a Denisovan dad.

Denny was an exciting, but not surprising find. We have indirect evidence of interbreeding: trace amounts of DNA from close "cousin" human groups have been found in Neanderthals, Denisovans, and modern humans. With the discovery of Denny, we have *direct* evidence of interbreeding between human groups. How often did it happen? That's a question to be answered, but Denny provides a hint. We've only known about Denisovans since 2008, and already we have a first-generation hybrid with another human group.

Modern Humans

The oldest modern humans (*Homo sapiens*) date to about 200,000 years ago in Ethiopia. Compared to other *Homo* groups, modern humans had similar musculature but a taller and lighter skeletal frame. Modern humans have a very big brain to body ratio. To house this big brain, the skull is high and thin, with a flat and vertical forehead. Unlike other *Homo* groups, modern human faces do not have heavy brow ridges. Our jaws are smaller, and we have smaller teeth.

Early modern humans left evidence of complex social groups, language, symbolism, art, jewelry, musical instruments, and burial practices. They used tools and fire, and cooked. They lived like contemporary hunter-gatherers.

Somewhere between eighty thousand and fifty thousand years ago, a single group of modern humans left Africa. How do we know this? Modern Africans are the most genetically diverse of all continental populations. The remainder of the world's population is far less genetically diverse than the modern African population. In other words, people living in the Americas, Europe, Asia, and Australia are more closely related to each other than are populations inside modern Africa. This evidence tells us all populations outside Africa descended from a small migrating founder group.[6] If earlier modern human groups left Africa, they were not successful and died out.

When modern humans moved out of Africa and trekked across the globe, they met some very ancient, but closely related cousins. The Neanderthals were spread out across Europe and western Asia and the Denisovans ranged from eastern Europe to eastern Asia.

All three groups (Neanderthals, Denisovans, modern humans) were genetically distinct. However, modern humans were closely related enough to the Neanderthals and the Denisovans to mate and have children.

How do we know this? People with European and Asian ancestry have trace amounts (1–4 percent) Neanderthal DNA in their genomes. Denisovan DNA is highest in the modern populations of southeast Asia and Oceania (4–6 percent).[7] Trace amounts of Denisovan DNA are also found in east Asian populations. Interestingly, people in sub-Saharan Africa have zero to almost zero Neanderthal or Denisovan DNA.[8]

Denny was a first-generation child of a Neanderthal mom and a Denisovan dad. We have yet to find a first-generation offspring of modern humans and Neanderthals or Denisovans, but they surely existed.

For several thousand years, our direct ancestors shared the planet with some of our close relatives (other hominins) who were not in our direct lineage—hominin aunts, uncles, and cousins, so to speak. We don't know the degree of the interactions, but we do know some of them were having children together.

But by thirty thousand years ago, modern humans—with our palette of DNA from other *Homo* groups—stood alone, the last remaining branch on the tangled human family tree.

We have some pretty good ideas why, but nothing definitive. Most likely, there was no single reason. At the time modern humans, Neanderthals, and Denisovans were sharing Europe and Asia, an ice age was deepening. Did encroaching glaciers overtake forest hunting lands and the Neanderthals failed to adapt? Did the superior brains, dexterity, speed, and language of modern humans allow them to survive while earlier human cousins did not? Did modern humans expose their human cousins to new diseases?

Was it a combination of the above along with the absorption of Neanderthal and Denisovan populations into modern human populations? Ongoing genetic studies are identifying genes inherited from Neanderthals and Denisovans in our human genome. Denisovan remains have been found at the edge of the Tibetan plateau and remarkably, the gene that allows modern Tibetans to live in a high altitude, low oxygen environment has been identified in Denisovan DNA.[9]

So, Who's the Missing Link?

Since Darwin published *The Descent of Man*, skeptics have demanded to see the missing link—the one and only, definitive half-monkey, half-human creature bridging ape to man.

To demand a single "missing link" is to misunderstand how evolution works. In human evolution (as well as in the evolution of all organisms), change does not occur in a straight-line, one-turning-into-the-other kind of process. Instead, evolution is a slow spreading and branching process, eventually resulting in greater and greater diversity. Over thousands and millions of years, many branches die out and part of the family tree is a dead end. Other branches survive and are the modern species we see today.

There is no one missing link fossil for humans. There are, in fact, at least twenty "links" exhibiting human traits, each found in or originating in Africa over the past five to six million years.[10] There is no way to know if a specific fossil species is a direct ancestor of modern humans. What we *can* learn from our close human cousins is the general trajectory of evolution leading to modern humans.

When we find a fossil, how do we know if it does or does not belong in the human family tree? Scientists look for skeletal traits unique to humans and not found in apes. The oldest fossils in the human tree are more ape-like, with a few human traits. Those in the mid-age fossil range (like Lucy) are mosaics of human and ape characteristics, with varying degrees of primitive and humanlike traits. The closer in age a fossil is to modern humans, the more humanlike traits we see and the fewer primitive traits.

Two of the oldest uniquely human skeletal traits are found in the skull. In humans, the opening for the spinal cord is forwardly placed, allowing for upright posture. Teeth are also a tell-tale sign—canine teeth are small in humans and large in chimps. Additionally, human teeth are arranged in an arch, chimp teeth are arranged in a rectangular form. Lucy had small canines and teeth arranged midway between a rectangle and an arch.

Over six million years, other uniquely human traits arose in our ancestors.[11] Human femurs (thigh bones) point inward, allowing upright walking. Humans have a short and broad pelvis, long legs, a long flexible waist, low shoulders, and a long flexible thumb. Humans have a large brain and the skull size and shape to accommodate it. Humans have small faces and a chin.

When scientists find fossils with some or all of these traits, they know they've found a human or a human relative, not an ape.

The Demand for a Missing Link

The demand for a missing link is as old as the first brand-new copy of Darwin's *On the Origin of Species.* Despite an abundance of physical fossil evidence and definitive molecular confirmation, the demand is still heard from creationists.

In analyzing any hominin fossil, creationists see no middle ground; fossils or fossilized footprints are either "definitely human" or "definitely ape." David Menton tells us how to analyze fossils from the creationist perspective:

> Knowing from Scripture that God didn't create any apemen, there are only three ways for the evolutionist to create one:
>
> - Combine ape fossil bones with human fossil bones and declare the two to be one individual—a real "apeman."
> - Emphasize certain humanlike qualities of fossilized ape bones, and with imagination upgrade apes to be more humanlike.
> - Emphasize certain apelike qualities of fossilized human bones, and with imagination downgrade humans to be more apelike.[12]

Any "evolutionist" interpretation of fossils is by default a combination of deceit and imagination.

Even among those who hold space for evolution, *human* evolution can be hard to swallow. Why would God create image-bearers using a cold, blind, random process, resulting in a being no different from a common animal?

A natural, unguided process does not, by default, equal a purposeless process. Here's Francis Collins:

> If God chose to create you and me as natural and spiritual beings, and decided to use the mechanism of evolution to accomplish that goal, I think that's incredibly

> elegant. And because God is outside of space and time, He knew what the outcome was going to be right at the beginning.[13]

The Missing Link—At Long Last

It's really hard to overstate the magnitude of the announcement. It has been called one of the "great feats of exploration in history." In 2003, the international Human Genome Project,[14] headed by Francis Collins, announced the complete mapping of the human genome—a map of all the genes of human beings. Here's the monumental announcement:

> [T]his Book of Life is actually at least three books. It's a history book: a narrative of the journey of our species through time. It's a shop manual: an incredibly detailed blueprint for building every human cell. And it's a transformative textbook of medicine: with insights that will give health care providers immense new powers to treat, prevent and cure disease. We are delighted by what we've already seen in these books. But we are also profoundly humbled by the privilege of turning the pages that describe the miracle of human life, written in the mysterious language of all the ages, the language of God.[15]

Already, mapping the human genome has fueled discovery of thousands of disease genes and allowed doctors to diagnose genetic risks like never before. Already we are tailoring cancer therapies to the specific type of cancer in a patient.

And in the Human Genome Project, the long-sought missing link tying chimpanzees to humans appears right before our very eyes.

Before the human genome was mapped, an abundance of fossil evidence indicated that humans share a common ancestor with the great apes (chimpanzees, gorillas, orangutans).

There was, however, a mysterious inconsistency at the chromosome level. Humans have twenty-three unique chromosomes, but we have two copies of each, one from biological mom, one from biological dad. Here's the problem: chimpanzees have twenty-four pairs of chromosomes, again one from chimp mom, one from chimp dad.

Was a chromosome pair lost at some point in the lineage leading to humans? Highly unlikely—the loss of so much genetic information would not be survivable.

What if, at some point in human evolution, two chromosomes fused? If a chromosome fusion occurred in human evolutionary history, what would we expect to see?

First, a little chromosome geography. All chromosomes have unique DNA landmarks at each end of the chromosome called *telomeres*. Telomeres are *only* found at the tips of chromosomes. There is also a unique DNA landmark called a *centromere* found only near the midpoint of a chromosome.

The Human Genome Project revealed that human chromosome 2 is like no other.[16] At each end of chromosome 2, right where you would expect them to be, are active telomeres. But *right in the middle* of the chromosome are two more telomere sequences. And there's more: human chromosome 2 has two centromeres instead of one.

But we're not done yet. The genes on human chromosome 2 line up and correspond almost exactly to chimpanzee chromosomes 12 and 13. At some point in the lineage leading to humans, two primate chromosomes fused, tip to tip. The molecular evidence is so strong we can identify the exact spot where the fusion occurred.[17]

Creationist response to the genetic evidence has primarily been formulated by staff writers at Answers in Genesis and the Institute for Creation Research.[18] Both organizations cite articles and research "debunking" the evidence for chromosome fusion. It is important to note, however, that *none* of the articles or the research cited have been published in peer-reviewed science journals—this despite the fact many of the articles from both organizations are written by Nathaniel Jeanson, holder of a PhD in cell biology from Harvard.[19]

Creationist articles debunking chromosome fusion are written with lots of technical terminology, with descriptors like "mythical," "alleged," and "premature" generously applied to the evidence—evidence that has been replicated and supported for almost two decades by international scientists. Again—and I cannot emphasize this enough—none of the articles refuting the genetic evidence are published in peer-reviewed science journals. All are published in creationist journals.

What About Adam and Eve?

About ten thousand to twelve thousand years ago, just after the last ice age, the world's cheetah population experienced a disaster. Was it an environmental event like a volcanic eruption? Was it a widespread disease? Were cheetahs hunted to the brink of extinction? We aren't sure. What we do know is this: a catastrophe of some sort reduced the world's cheetah population to only seven to twelve individuals. All cheetahs living today are descended from that tiny population of a dozen or less.

Modern cheetahs are essentially identical and highly inbred. A cheetah can accept a skin graft or a kidney transplant from almost any other cheetah without rejecting the donated tissue or organ.

And cheetahs aren't the only ones. The Tasmanian devil, one of Australia's quirky marsupial mammals, is in grave danger of extinction. Like cheetahs, Tasmanian devils are essentially genetically identical, due to one or more population-reducing events in their history. A particularly gruesome form of contagious facial cancer is decimating the modern population, transferred from devil to devil by bites around the face and neck. Normally, the body recognizes cancer cells as "not self" and the immune system kicks in to stop the invaders. But because they are so genetically identical, Tasmanian devils can "donate" their tumor cells to any other devil, and the cells will not be rejected.

If all humans descended from one man and one woman in the last six to ten thousand years, what kinds of things would we expect to see? Blood donations would be simple—no need for cross-matching. Organ transplants? Piece of cake. No need to match donors with recipients, and no need for anti-rejection drugs. If all humans descend from a population of two, humans would be so genetically identical any sort of transplant would be a breeze.

But that is not what we see. The broad genetic diversity in modern humans could only arise from a large founding population.

Copying errors (mutations) occur as DNA copies itself in normal growth and cell division. Over generations and generations, these changes accumulate and result in different versions of a gene. For example, there is a functioning version of the hemoglobin gene and a sickle-cell disease version of the hemoglobin gene. Often there are many versions of a gene in a population; the gene for human blood type comes in three versions: A, B, and O.

We know the rate of human gene mutation: about 100–200 new mutations in each generation.[20] Thanks to the Human Genome Project, we can look at individual genes and count how many versions of that gene exist in the human population. In studies tackling this math problem from several different angles, the answer is always the same: modern humans descend from an ancestral population of about ten thousand individuals who lived around 150,000 years ago.[21]

Suddenly, we have a problem with Adam and Eve. We have the genes, we know the rates, we can do the math. Can we make the mathematical evidence fit Genesis?

Again, Nathaniel Jeanson is the primary spokesperson for this dilemma. Jeanson does not deny the math or the calculations or even the answer, he simply disagrees with the conclusion. His explanation is straightforward: designed genetic variability.[22] According to Jeanson, the genetic diversity we see in humans, and in all life for that matter, is not what it appears to be. Instead, God created humans and all life with the appearance of a genetic history.

Were Adam and Eve real, actual, and historical people? Or, is the story of Adam and Eve intended to communicate something other than literal history? These are questions for theology, not science. Many biblical scholars have tackled this topic and are well worth your investigation.[23]

Maybe Adam and Eve were actual, historical people, maybe they were not. Science does not address that question any more than science addresses the historicity of Abraham, David, or Esther.

What Adam and Eve *cannot* be, however, are the literal, genetic ancestors of all humanity.

12

Leaving Creationism (Without Leaving God)

Old, frail, and at the end of a life of privilege, fame, and popularity, the great astronomer and telescope whiz Galileo Galilei found himself in Rome, on trial before the church Inquisition. The politics of Inquisition court proceedings were complicated in 1633 but suffice it to say threats of torture were definitely allowed. Galileo was not martyr-material, so he took the seventeenth-century equivalent of a plea bargain and lived out his life in "villa arrest." His crime? Belief in a sun-centered solar system.

Galileo did not invent the telescope, but he built the best one of his day. Observations using his state-of-the art telescope convinced Galileo that the sun (not the earth) was actually the center of the solar system and that all planets, including the earth, travelled around the sun. For decades, Galileo published and publicized his sun-centered position. And Galileo wasn't the only one; others within the jurisdiction of the church believed the same.

What then, put Galileo in the crosshairs of the Inquisition? First of all, his personality didn't help. As popular as he was and as adept as he was in accumulating friends, Galileo also excelled at making enemies. Galileo had the habit of making his academic points with insults and name-calling, regardless of the status of his target.

The real problem, however, was not with science. The real problem was with theology. The Catholic Church was in a defensive position, post-Reformation. Similar to the Protestants of the day, the Catholic Church claimed its practices were based on a "plain reading of Scripture."[1] The points of contention were literal readings of several Old Testament verses: "the world also is established, that it cannot be moved"; "who laid the foundations of the earth, that it should not be removed forever"; and in the book of Joshua "the sun stood still."[2]

Not one to back down, Galileo decided to add a theological component to his argument. Galileo made his point: the purpose of Scripture is to teach men how to go to heaven, not how the heavens go.[3] To say church leaders were displeased with Galileo's amateur foray into theology is an understatement. Stay in your lane, Galileo. Stay in your lane.

How dare Galileo deny the plain meaning of Scripture? How dare Galileo question long-standing beliefs, passed down over millennia and accepted by the church fathers?

In reality, science wasn't the problem. It was theology. If the earth is not the center and is just one of innumerable planets in the universe, the earth is not special to God, and therefore man holds no special place in creation. Man is no longer the apple of God's eye and an entire belief system falls apart. The church would not allow it.

Galileo was humiliated and broken. He died without exoneration.

But evidence accumulated, books were written, and telescopes abounded. Isaac Newton burst onto the scene. Before the end of the seventeenth century, the church had no problem, scientifically or theologically, with a sun-centered solar system. In the face of unequivocal scientific evidence, Christians changed long-held interpretations of Scripture. The earth-not-moving Scriptures still speak truth, but the truth they speak is not literal science truth.

Modern arguments against evolution sound very much like the theological arguments against a sun-centered solar system. If humans are just another branch in the evolutionary tree, we aren't special to God. If we don't have a real Adam, there can be no "fall," and without the fall, there's no need for Jesus. If we can't believe Genesis, we can't believe any of the Bible. All is lost.

Biblical Meteorology

I live in an area of Texas where spring or fall days can give us all four seasons in one 24-hour period. The heart of summer is a different story, however, when the dry days of July and August and non-stop 100°F temperatures stretch into long weeks, even months. Social media feeds are scrolls of memes, all with the same message: pray for rain. Long before social media, prayers for rain were mainstays in our homes and churches. And when the rain comes, we credit God and give thanks.

Yet, no one wants to forbid teaching the water cycle in public schools. No special interest groups are pressuring science teachers to "teach the controversy" of the water cycle or to present alternative explanations for rain. This, despite the fact Scripture explicitly states that rain, snow, and hail are kept in storehouses and from these storehouses God pours out rain on the earth. Likewise, Scripture teaches wind is kept in a container and sent to earth by God. God also physically churns the clouds in the sky.[4]

No one debates the water cycle because it contradicts what the Bible says about precipitation. No one is advocating a "physical storehouses model" to make the science fit Genesis. No one denies the science of air pressure and temperature differences in the creation of wind. No one is insisting on biblical meteorology.

Weather is physical science, but what about the creation of new life? Scripture plainly states that God directly forms us ("knits" us) in our mother's womb.[5] Yet, no one denies the nine-month natural developmental process resulting in a baby.

Water evaporation, condensation, and the tilt and rotation of the earth on its axis result in rain, seasons, day, and night. Parents thank God for the gift of a child with the complete understanding this gift is a nine-month, fertilized-egg-to-baby natural process. The world functions according to observable and predictable natural processes, yet we still credit God.

Biology, geology, and astrophysics are the only modern sciences that are suspect, and only when applied to questions of origins. As a science educator, it troubles me to see observable evidence ignored in lieu of a "what could have happened" explanation as a way to force-fit modern science into Genesis. As a Christian, I fear the damage done by such a force-fit.

When You Don't Know the (Science) Answer, the Answer Is Always God

The great Isaac Newton did not have all the answers. The inventor of calculus and discoverer of the laws of motion and gravity was stumped when it came to planetary orbits. No one, including Newton, understood the details: Why did planets all travel in the same direction? Why are all planets located in the same plane? And why is it all so consistent?

Newton's answer? God. Lacking a natural explanation, Newton concluded God simply designed it to run this way.[6] And yet, a century after Newton, better tools and better math provided the answer—the science answer.

The rallying cry of the creationist/intelligent design movement is the unexplainable. Wherever there is a gap in what we know, wherever there is a piece of the puzzle we haven't found, wherever there is a step or a structure yet to be discovered, the default answer is God. Unexplained steps in biochemical processes are chalked up to a design so complex, nature could never produce it.

The go-to example in the category of "what we don't know" is the origin of life. In a letter to a friend, Charles Darwin speculated about a "warm little pond."[7] Darwin's premodern science, shot-in-the-dark hypothesis fuels much mocking and derision of evolution theory. "So we all just crawled out of the mud and goo? One day an amoeba crawled out of the muck and decided to become a fish?" You get the idea, and I'm sure you've heard some version of this argument.

Just because we don't have a science explanation for how life emerged today does not mean we won't have one tomorrow. It is tempting to put God in our gaps of knowledge: since we don't know how life began, it must have been a special miraculous intervention by God. Then what happens if tomorrow's headlines announce scientists have discovered how life began? Is all lost for believers in God?

Hardly.

If (or when) we discover how the first molecules of DNA formed, or if we discover how a membrane organized around a molecule of DNA, or if we discover how membrane-enclosed DNA replicated and formed new cells, we will not have disproved God. Instead, we will have discovered the mechanism by which God brought life about on earth.

"I've learned to praise God for *what is glorious in his creation*, even if it isn't miraculous," writes Deborah Haarsma, "God is just as present in the regular workings of nature as he is in supernatural acts."[8]

After a lifetime of church-going, I've heard countless songs praising God in terms of the created world—the stars, the rolling thunder, "thy power throughout the universe displayed."[9] A few years ago, our worship leader introduced Hillsong United's "So Will I (100 Billion X)." The melody was beautiful and ethereal, but the lyrics in the first verse stopped me in my tracks.

We are singing about the universe, old and immense. We are singing about planets forming in the wake of stars. This is real science.

Ok, Hillsong. You have my attention. And with the next verse, I am undone.

God speaks and nature listens. *Science* listens.

And not in a micromanaging, every-step-a-miracle way, but through the natural laws and processes of science.

Through *evolution.*

No mental gymnastics needed to make science fit into a literal Genesis. Evolution is the process responsible for the brilliant diversity of life on our planet.

And we praise God for it. In a song. In church.

Putting Up Walls

Some years ago, I was invited by our (then) minister into a conversation with a couple in our church's English as a second language program. The husband was an international student doing post-doctoral studies in sciences at a local university. The wife was eager to explore Christianity, but her husband was far more skeptical. I was told he had lots of questions. Hoping our apologetics skills were up to snuff, my husband and I met them for dinner.

After introductions and small talk, we dove right in. And wow, did we ever. No easing into the shallows first with this guy. His first question was blunt and straight to the point: why don't Christians accept science?

In just a few sentences, we explained that there was no conflict between Christianity and science. We briefly told him about preeminent scientists who were practicing Christians. His body language immediately changed. At that point, he couldn't get his questions out fast enough! One after another.

But the questions weren't about reconciling science and faith. Every question was about Jesus. He was drawn to Jesus, but science-denial was a barrier he just couldn't get around. Once the wall was down, he was freed.

My post-doc science friend is not the only one.

The Barna Group has tracked trends in the church for

decades. The current picture in America is not a pretty one. Church attendance and religious affiliation are waning and the number of young adults identifying as atheists is growing.[10] The fastest growing group is those who were active churchgoers in the past but who are active no longer. Barna calls this group the "de-churched."[11]

The reasons for de-churching are varied and nuanced, but Barna has identified six consistent themes among the dropouts. Among the consistent six? "The church is antagonistic to science."[12] Young adult dropouts (and older ones as well) believe the church is out of step with modern science and even *anti*-science. They are particularly turned off by the creation versus evolution debate.

Before a live-streamed audience of 500,000, Bill Nye (the Science Guy) and Ken Ham (the Answers in Genesis guy) debated the merits of creationism and evolution. As of this writing, the Ham-Nye debate has been viewed almost eight million times.[13] From the center of his six-thousand-year-old universe, Ham drew a line in the sand: choose science or choose God. There can be no compromise. And across the internet, like the residents of Whoville yelling and whooping with big banging noises trying to make themselves heard to the outside world, thousands of Christians were shouting: "It doesn't have to be that way!"

What Questions Are We Missing?

Andrew Root is a seminary professor whose research interest is youth ministry. His Science and Youth Ministry project was an exhaustive examination of science and faith in the youth ministries of American churches.[14]

While the focus of his research was youth ministry, Root and his team exposed a connection between science and faith with implications for all of us.

As counterintuitive as it seems, Root's research found inclusion of science conversations in youth ministry actually stimulated conversations about God. In other words, conversations about what we can know factually lead to conversations about what we must accept by faith. By acknowledging the validity of science, Root believes we can address a much larger issue: what does it mean to have faith in a modern world of facts and evidence?

What important science and faith discussions are we missing? In an evidenced-based world, what questions cannot be answered by science? What could Christians bring to the science table if we weren't preoccupied with making science fit a literal reading of Genesis?

Charles Darwin appeared to have lost his religious faith by the end of his life. But contrary to popular lore, Darwin's loss of faith was most likely *not* due to his acceptance of evolution. Darwin struggled with questions of evil, violence, and death in the natural world, including the devastating death of his young daughter—weighty questions for all of us. Christians are uniquely equipped to respond to these kinds of questions but are too often bogged down in defending a "science" that isn't science.

What Is the Cost?

Extreme voices in the science and faith conversation (think Ken Ham and Richard Dawkins) draw honest seekers to the edges. Anything between the edges of extremes is defined as either a compromise of faith or a compromise of intellect. What are people of faith to do when they want to love God with heart, soul, *and* mind? Ignore science? Pretend it isn't so? Is there an option honoring both God and science?

The journey to accepting evolution is often incremental and sometimes includes a time of closeted acceptance of evolution. *How I Changed My Mind About Evolution* is a collection of twenty-five short memoirs—firsthand accounts by scientists, pastors, biblical scholars, and theologians.[15] The backgrounds and stories vary, but common themes wind their way through the memoirs.

One theme, however, is pervasive. In some form or other, a realization of personal intellectual dishonesty percolates through most of these faith memoirs. The mental gymnastics required in order to make evidence fit a young earth, a literal Genesis, or the claims of the intelligent design movement eventually became harder than accepting evolution.

Consider: A universe less than ten thousand years old is not what it appears to be. We know the speed of light and the distance of the stars and we can do the math. But if the universe is only a few thousand years old, the light you see from countless stars and galaxies hasn't really travelled thousands and millions of light years. Perhaps the universe was created "full-grown" and starlight was created in transit. Or, we can change the rules: "the speed of light was faster at the beginning."[16] Either way, starlight was created by God to look ancient, but in reality, it is not. The night sky is a breathtakingly beautiful work of science fiction.

Consider: An earth less than 10,000 years old is not what it appears to be. Somehow, a young earth has almost a million years of sediment and ice accumulation. Somehow, radioactive rocks are found in various degrees of decay, despite an instantaneous creation. Even though all metrics used to measure the age of the earth point to billions of years, this is simply designed appearance of age.

Consider: Creation of all living things instantaneously means the fossil record is not what it appears to be. The chronology of life as seen in fossil layers, from simple cells to complex four-limbed creatures, is simply an accident, an artifact of receding flood waters.

Consider: Intelligent design demands living things appear to have an ancestral history that they in fact, do not. Body architecture that appears ancestral is actually the result of a designer repeating design patterns. Nonfunctional genes found scattered throughout the DNA of an organism while functional versions of these same genes are found in the DNA of other organisms are likewise a design choice. Intelligent design demands a designer who intentionally inserts broken genes into functioning DNA. Intelligent design demands a designer who designs organisms that look for all the world like they share common ancestry, when in fact, they do not.

Consider: Belief in a young earth and/or special creation (including intelligent design) has particular implications for science research. If properties of physics and chemistry are not what they appear to be in the case of origins, how can we trust physical properties in any other context? If one biochemical process or cellular structure is so complex it demands design, why not all processes and structures?

Consider: If the intricacies of life are all choices made by a designer, according to a design known only to the designer, how is anything knowable? If the complexities of life cannot be explained by natural processes, why research? The strength of any science theory is its ability to predict new knowledge. Choices made by a supernatural designer, known only to the designer, are a scientific dead end. Nothing is predictable.

Force-fitting science into Genesis comes at an intellectual cost, and creationists must decide if the cost is too high.

There is no one discovery, no one piece of evidence, no singular "a-ha!" moment that enthrones evolution as the theory that explains it all. Instead, confidence in evolution comes from a convergence of evidence from diverse fields across science: anatomy, physiology, ecology, genetics, biochemistry, molecular biology, geology, paleontology, physics, astronomy, and chemistry.

A literal Genesis means a stand against the vast majority of modern science and scientists.[17] A literal Genesis means a stand against the science trusted for medical care, disease research, agriculture, aviation, engineering, and energy. A literal Genesis means mentally segregating the science explaining origins from the same science supporting our modern lives. If creationism is true (including young, old, and intelligent design versions), modern science collapses.

In addition to an intellectual cost, there is also a faith cost. A literal reading of Genesis requires an honest reflection about the nature of God:

> The heavens declare the glory of God;
> the skies proclaim the work of his hands.
> Day after day they pour forth speech;
> night after night they reveal knowledge.
>
> Psalm 19:1–2

If God's creation "reveals knowledge," is the revealed knowledge trustworthy? Is it consistent with God's nature to fill creation with red herrings? Is it consistent with God's nature to create a deceptive world, a world not as it appears?

Is it consistent with God's nature to mislead us?

When I first began to seriously study what it means to reconcile science and faith, I learned this Scripture: "God is not human, that he should lie."[18] The heavens reveal knowledge, and God does not lie.

What Do We Do with Genesis?

What then, are committed people of faith to do when they want to love God with heart, soul, and mind? Ignore science? Pretend it isn't so? Or, revisit the way we read Genesis?

Put on your headphones and listen to "Home" by Phillip Phillips. Phillips contrasts a physical place (a house) to the home he wants to build, a home of love, belonging, and safety. It's a well-loved theme: sheetrock and nails make a house, but they don't make a home.

Old Testament scholar John Walton applies the metaphor to Genesis.[19] When scientists talk about origins of the universe and life, they talk about the materials, the processes, the blueprints. Scientists study the house. God built the house, Walton says, but Genesis is not a house story. Genesis is a home story.

What does a science-accepting person do with Genesis? Ignore it? Not at all. Honoring Genesis means recognizing its genre. Honoring Genesis means listening to its ancient voice. It's a fascinating study; some of the best scholarship is being done by John Walton, Scot McKnight, and Pete Enns.[20]

Trying to force a modern understanding of science into an ancient document misses lots of boats. Not only do we miss the message intended by the original authors and compilers, we also force the Bible to be something it is not—a scientifically accurate natural history of the earth. When we read Genesis, we don't learn about modern science, but we *do* learn about God.

For me, as a Christian, the Bible has authority because it testifies to Jesus. The Bible has authority because it bears witness to people trying to live as God's people. The Bible has authority in how I live my life. But the Bible is not an authority about the facts of modern science. It was never meant to be. The Bible is telling a different story. The Bible gives us the answers to the *who and why* of creation; science answers the *how and when.*

Please don't confuse science with the ideas of *scientism* or *materialism*. Both are generally understood to mean all reality can be explained in material or physical terms. Science does not answer all the questions. The most important questions humans ask cannot be answered with science.

It may be a struggle; it may take a journey. The best of minds wrestle with Genesis and modern science. Francis Collins is a world leader in genetics and medicine. He left atheism as a young doctor, followed Christ, and publicly testifies about his faith. Yet, he acknowledges the discordance:

> The evidence supporting the idea that all living things are descended from a common ancestor is truly over-

> whelming. I would not necessarily wish that to be so, as a Bible-believing Christian. But it is so. It does not serve faith well to try to deny that.[21]

Evolution is elegant. Evolution is creative. Evolution produced a creation that continues to create. Evolution is not inherently godless. Evolution does not diminish God as creator and sustainer of all there is. Nothing about evolution excludes God as the one in whom "we live and move and have our being."[22]

My faith tells me God ordained it all.

Our creation, in its material and natural form, function, and origin, is what God, in God's wisdom, gave us.

Notes

Chapter 1

1. The "Gish gallop" is named for creationist Duane Gish for his propensity to overwhelm his debate opponents with mountains of questions and assertions that could not be answered or countered in the allotted timeframe. The term is used beyond the context of a creationism-evolution debate whenever someone tries to win an argument by overwhelming an opponent with talking points.

Chapter 2

1. Michael Hoskin, ed., *The Cambridge Illustrated History of Astronomy* (Cambridge: Cambridge University Press, 1997), 75.

2. Cuvier was between a rock and a hard place. As the preeminent paleontologist of his day, he knew fossils told a story of change over time, but as a religious man, he staunchly defended the Genesis creation story. Cuvier proposed an unknown creation prior to the Genesis creation as the solution to his dilemma. More and more geologic formations were soon identified, and others followed Cuvier's lead. As a result, a complicated cycle of separate creations and floods was suggested in order to reconcile the science with Genesis.

3. Donald R. Prothero, *Evolution: What the Fossils Say and Why It Matters* (New York: Columbia University Press, 2017), 59–61.

4. "Kids Who Love Dinosaurs Have Much Higher Intelligence, Study Finds," *The Manc*, April 5, 2019, https://themanc.com/news/kids-who-love-dinosaurs-have-much-higher-intelligence-study-finds/.

5. Ashley Powers, "Adam, Eve, and T. Rex," *Los Angeles Times*, August 27, 2005, https://www.latimes.com/archives/la-xpm-2005-aug-27-me-dinosaurs27-story.html.

6. For example, see Kyle Butt and Eric Lyons, *Dinosaurs Unleashed: The True Story about Dinosaurs and Humans* (Montgomery: Apologetics Press, 2004).

7. Job 40:15; Job 41:1; Psalm 91:13 (KJV).

8. See Brian Thomas, "Were Dinosaurs on Noah's Ark?," Creation Q & A, Institute for Creation Research, January 29, 2016, https://www.icr.org/article/were-dinosaurs-noahs-ark; and Buddy Davis, "Dinosaurs on the Ark," Answers in Genesis, February 24, 2010, https://answersingenesis.org/dinosaurs/humans/dinosaurs-on-the-ark/.

9. *Patagotitan mayorum*, a titanosaur, was 120 feet long and weighed 69 tons; see Shaena Montanari, "New Dinosaur Species Was Largest Animal Ever to Walk the Earth," *National Geographic*, August 9, 2017, https://www.nationalgeographic.com/news/2017/08/largest-dinosaur-ever-titanosaur-fossil-patagotitan-science/.

10. Megan Brenan, "40% of Americans Believe in Creationism," *Gallup News*, July 26, 2019, https://news.gallup.com/poll/261680/americans-believe-creationism.aspx.

11. April Maskiewicz Cordero, interview with Jim Stump and Kathryn Applegate, *Language of God*, podcast audio, January 22, 2020, https://biologos.org/podcast-episodes/april-cordero-teaching-in-the-tension.

12. Andrew Root, David Wood, and Tony Jones, "Youth Ministry & Science: A Templeton Planning Grant," January 2015.

13. The focus of AiG's college prep book *Already Compromised* (Ken Ham and Greg Hall [Green Forest, AR: Master Books, 2011]; https://answersingenesis.org/answers/books/already-compromised/) is Christian colleges. Parents are warned: you are footing the bill for professors to destroy the faith of your children and teach them to believe in evolution. The Discovery Institute's *The College Student's Back to School Guide to Intelligent Design* (2009; http://www.arn.org/docs/BacktoSchoolGuide_Sept2009_FN-1.pdf) suggests responses to professors' "misinformation."

14. Jason Lisle, *How to Survive Secular College*, Answers in Genesis video, https://answersingenesis.org/college/.

15. Read the transcript of a 90-minute discussion between Richard Dawkins and Francis Collins: David Van Biema, "God vs. Science," *Time*, November 5, 2006, http://content.time.com/time/magazine/article/0,9171,1555132-1,00.html.

16. *Young Sheldon*, season 1, episode 3, "Poker, Faith and Eggs," aired November 9, 2017, on CBS.

17. *God's Not Dead*, directed by Harold Cronk (Pure Flix, 2014).

Chapter 3

1. Bruce Y. Lee, "Nobel Prize Winner Frances Arnold Retracts Paper, Here Is the Reaction," *Forbes*, January 5, 2020, https://www.forbes.com/sites/brucelee/2020/01/05/nobel-prize-winner-frances-arnold-retracts-paper-here-is-the-reaction/#71d5683032c1.

2. Michael Greshko, "This Dinosaur Is the 'Most Impressive Fossil' We've Ever Seen," *National Geographic*, May 14, 2017,

https://www.nationalgeographic.com/photography/proof/2017/05/nodosaur-fossil-discovery-science-photography/.

3. "Well-Preserved 'Mona Lisa' of Dinosaurs Unearthed in Canada," *news.com.au*, August 5, 2017, https://www.news.com.au/technology/science/animals/wellpreserved-mona-lisa-of-dinosaurs-unearthed-in-canada/news-story/c94efecbf530683b5e0480614af19df1.

4. Charles Darwin, *On the Origin of Species*, originally published November 24, 1859.

5. "Evolve or Die: Strong Selective Pressures Repurpose Gene Function," in *Biotechnology Discoveries and Applications* (Huntsville, AL: HudsonAlpha Institute for Biotechnology, 2016), 10, https://s3.amazonaws.com/hudsonalpha/wp-content/uploads/2015/11/02223430/Guidebook-for-web.pdf. See also Ruth Williams, "Evolutionary Rewiring," *The Scientist*, February 26, https://www.the-scientist.com/daily-news/evolutionary-rewiring-35878.

Chapter 4

1. Answers in Genesis, https://answersingenesis.org.

2. The Institute for Creation Research, https://www.icr.org.

3. Eric Hovind, https://creationtoday.org/eric-hovind/.

4. *We Believe in Dinosaurs* is a documentary about Answers in Genesis's Ark Encounter museum, from conception to opening day. The film features the creationists behind the project and highlights opposition from the local science community. https://www.webelieveindinosaurs.net.

5. Apologetics Press, http://apologeticspress.org.

6. *Dennis Venema and Georgia Purdom LeTourneau University April 11 Part 3*, posted April 14, 2014, https://www.youtube.com/watch?v=fb7QNFdW7Cc&feature=youtu.be.

7. David Masci, "For Darwin Day, 6 Facts about the Evolu-

tion Debate," Pew Research Center, February 11, 2019, https://www.pewresearch.org/fact-tank/2019/02/11/darwin-day/.

8. Todd Wood, "The Truth about Evolution," Todd's Blog, September 30, 2009, http://toddcwood.blogspot.com/2009/09/truth-about-evolution.html.

9. See for example, Ken Ham, "What's the Best 'Proof' of Creation?," Answers in Genesis, March 18, 2010, https://answersingenesis.org/evidence-for-creation/whats-the-best-proof-of-creation/ and John D. Morris, "Does Science Conflict With the Bible?," Institute for Creation Research, November 1, 1997, https://www.icr.org/article/1173.

10. *We Believe in Dinosaurs*, https://www.webelieveindinosaurs.net.

11. Reasons to Believe, https://reasons.org.

12. Hugh Ross, *Creation as Science: A Testable Model Approach to End the Creation/Evolution Wars* (Colorado Springs: NavPress, 2001).

13. Does God Exist?, https://www.doesgodexist.org.

14. Fazale Rana, "No Joke: New Pseudogene Function Smiles on the Case for Creation," Reasons to Believe, April 1, 2020, https://reasons.org/explore/blogs/the-cells-design/read/the-cells-design/2020/04/01/no-joke-new-pseudogene-function-smiles-on-the-case-for-creation.

15. Jeff Zweerink, "A Best of TNRTB: General Revelation Affirms Scripture Accounts," Reasons to Believe, December 31, 2010, https://reasons.org/explore/blogs/todays-new-reason-to-believe/read/tnrtb/2010/12/31/a-best-of-tnrtb-general-revelation-affirms-scripture-accounts.

16. Ronald L. Numbers, *The Creationists* (Cambridge: Harvard University Press, 1992), 7.

17. Numbers, *The Creationists*, 60.

18. Discovery Institute, https://www.discovery.org.

19. Josh McDowell and Sean McDowell, *Evidence That Demands a Verdict* (Nashville: Thomas Nelson, 2017).

20. *Think Biblically* is a podcast hosted by Scott Rae and Sean McDowell of Biola University: https://www.biola.edu/blogs/think-biblically/about.

21. The Center for Science and Culture, "What Is the Science Behind Intelligent Design?," Discovery Institute, May 1, 2009, https://www.discovery.org/a/9761/.

22. BioLogos, https://biologos.org.

23. The Templeton Prize honors individuals whose exemplary achievements advance the Foundation's vision of "harnessing the power of the sciences to explore the deepest questions of the universe and humankind's place and purpose within it." Francis Collins was the 2020 recipient; https://www.templetonprize.org.

24. Francis S. Collins, *The Language of God* (New York: Free Press, 2006).

25. Kenneth R. Miller, *Finding Darwin's God* (New York: Cliff Street Books, 1999) and *Only a Theory: Evolution and the Battle for America's Soul* (New York: Viking, 2008).

26. Dennis R. Venema and Scot McKnight, *Adam and the Genome* (Grand Rapids: Brazos Press, 2017).

27. See "What Is Evolutionary Creation?," Common Questions, BioLogos, https://biologos.org/common-questions/what-is-evolutionary-creation.

28. "Are Science and Christianity at War?," Common Questions, BioLogos, https://biologos.org/common-questions/are-science-and-christianity-at-war.

29. Since the early twentieth century, naturalism has been understood as a rejection of the supernatural and the superiority of science to explain all things. See "Naturalism," *Stanford Encyclopedia of Philosophy*, rev. March 31, 2020, https://plato.stanford.edu/entries/naturalism/.

30. Jerry Coyne, "Yes, There Is a War Between Science and Religion," The Conversation, December 21, 2018, https://theconversation.com/yes-there-is-a-war-between-science-and-religion-108002.

31. Richard Dawkins, *The God Delusion* (New York: Houghton Mifflin, 2006).

32. Dan Cray, "God vs. Science, Richard Dawkins and Francis Collins Interviewed by D. Cray," *Time International*, 2006.

33. Mark McGivern, "Parents' Outrage as Extremist US Religious Sect Hand Out Creationist Books and Preach to Kids at Scottish School," *Daily Record*, September 6, 2013, https://www.dailyrecord.co.uk/news/scottish-news/parents-outrage-extremist-religious-sect-2254926.

34. RJS, "Chew It Through Afresh", *Jesus Creed* (blog), July 16, 2013, https://www.patheos.com/blogs/jesuscreed/2013/07/16/chew-it-through-afresh-rjs/.

35. Peter Enns, in his book *The Bible Tells Me So* (New York: Harper Collins, 2014), discusses this concept in depth.

Chapter 5

1. Alaina G. Levine, "Holmdel Horn Antenna, Holmdel, New Jersey: The Large Horn Antenna and the Discovery of Cosmic Microwave Background Radiation," APS Physics, 2009, https://www.aps.org/programs/outreach/history/historicsites/penziaswilson.cfm.

2. The universe is wrapped in a blanket of microwave radiation left over from the big bang. See NASA website https://www.nasa.gov/vision/universe/starsgalaxies/cobe_background.html; also see Karl Tate, "Cosmic Microwave Background: Big Bang Relic Explained (Infographic)," Space.com, April 3, 2013, https://www.space.com/20330-cosmic-microwave-background-explained-infographic.html.

3. See "How Do We Measure the Distance to Stars?," *Scientific American*, December 2, 2014, https://www.scientificamerican.com/video/how-do-we-measure-the-distance-to-s2013-08-06/; and also "Parallax and Distance Measurement," Las Cumbres Observatory website, https://lco.global/spacebook/distance/parallax-and-distance-measurement/.

4. See "Gaia," Science and Technology, European Space Agency, https://sci.esa.int/web/gaia/.

5. Measure the speed of light for yourself at home using a microwave and chocolate! Colin Schultz , "There's an Easy (and Tasty) Way to Measure the Speed of Light at Home," Smithsonian Magazine, August 4, 2014, https://www.smithsonianmag.com/smart-news/theres-easy-and-tasty-way-measure-speed-light-home-180952245/.

6. Read about the Doppler effect on the Exploratorium's website: Pearl Tesler, "Physics 101: Redshift and the Expanding Universe," https://www.exploratorium.edu/origins/hubble/tools/doppler.html.

7. See NASA website: "The Farthest Visible Reaches of Space," The Cosmic Distance Scale, Imagine the Universe, Goddard Space Flight Center, https://imagine.gsfc.nasa.gov/features/cosmic/farthest_info.html.

8. United States Department of Agriculture Forest Service, "Ancient Bristlecone Pine Natural History," https://www.fs.usda.gov/detail/inyo/learning/nature-science/?cid=stelprdb5138621.

9. Michael Slezak, "Muddy Lake Bed Holds Radiocarbon 'Rosetta Stone,'" *New Scientist*, October 18, 2012, https://www.newscientist.com/article/dn22396-muddy-lake-bed-holds-radiocarbon-rosetta-stone/.

10. Read about ice cores on the NASA Climate website: Jessica Stoller-Conrad, "Core Questions: An Introduction to Ice Cores," August 14, 2017, https://climate.nasa.gov/news

/2616/core-questions-an-introduction-to-ice-cores/; see also "About Ice Cores," NSF Ice Core Facility, https://icecores.org/about-ice-cores.

11. Read about radioactive decay: Paul Flowers, Edward J. Neth, William R. Robinson, PhD, Klaus Theopold, Richard Langley, *Chemistry: Atoms First 2e* (Houston, Texas: OpenStax, 2019), https://openstax.org/books/chemistry-atoms-first-2e/pages/20-3-radioactive-decay.

12. "Jack Hills Zircon: Scientists Discover Oldest-Known Fragment of Earth," *Sci-News.com*, February 24, 2014, http://www.sci-news.com/geology/science-jack-hills-zircon-oldest-known-fragment-earth-01779.html.

13. Sarah Zielinski, "Earth's Magnetic Field Is at Least Four Billion Years Old," *Smithsonian Magazine*, July 30, 2015, https://www.smithsonianmag.com/science-nature/earths-magnetic-field-least-four-billion-years-old-180956114/; Nadia Drake, "No, We're Not All Doomed by Earth's Magnetic Field Flip," *National Geographic*, January 31, 2018, https://www.nationalgeographic.com/news/2018/01/earth-magnetic-field-flip-north-south-poles-science/; Lisa Grossman, "Earth's Magnetic Field is 3.5 Billion Years Old," *WIRED*, March 5, 2010, https://www.wired.com/2010/03/earths-magnetic-field-is-35-billion-years-old/; "2012: Magnetic Pole Reversal Happens All the (Geologic) Time," NASA, November 30, 2011, https://www.nasa.gov/topics/earth/features/2012-poleReversal.html.

14. Megan Brenan, "40% of Americans Believe in Creationism," *Gallup News*, July 26, 2019, https://news.gallup.com/poll/261680/americans-believe-creationism.aspx.

15. Calvin Smith, "Dinosaur Soft Tissue," *Creation Ministries International*, February 28, 2019, https://creation.com/dinosaur-soft-tissue.

16. Watch this short video by Stated Clearly for a quick and understandable summary of Mary Schweitzer's work: "Soft

Tissue Found Inside a Dinosaur Bone!," September 19, 2017, https://youtu.be/bSaOS7erEOk.

17. Helen Fields, "Dinosaur Shocker," Smithsonian Magazine, May 2006, https://www.smithsonianmag.com/science-nature/dinosaur-shocker-115306469/.

18. Francis B. Harrold, "Past Imperfect: Scientific Creationism and Prehistoric Archeology," *Creation/Evolution Journal* 10, no. 1, (Summer 1990), https://ncse.ngo/past-imperfect-scientific-creationism-and-prehistoric-archeology.

Chapter 6

1. See Francisco Del Rio Sanchez, "Discovering Gilgamesh, the World's First Action Hero," *National Geographic*, https://www.nationalgeographic.com/history/magazine/2018/01-02/history-gilgamesh-epic-discovery/; see also David Damrosch, "Epic Hero," *Smithsonian Magazine*, https://www.smithsonianmag.com/history/epic-hero-153362976/.

2. Irving Finkel, "Was the Ark Round? A Babylonian Description Discovered," *The British Museum Blog*, January 24, 2014, https://blog.britishmuseum.org/was-the-ark-round-a-babylonian-description-discovered/; see also Irving Finkel's presentation at the Oriental Institute, *The Ark before Noah: A Great Adventure*, July 20, 2016, https://youtu.be/s_fkpZSnz2I.

3. William Ryan and Walter Pitman, *Noah's Flood: The Scientific Discoveries about the Event That Changed History* (New York: Simon & Shuster, 1998).

4. David MacDonald, "The Flood: Mesopotamian Archaeological Evidence," *Creation/Evolution Journal* 8, no. 2 (Spring 1988), https://ncse.ngo/flood-mesopotamian-archaeological-evidence.

5. John C. Whitcomb and Henry M. Morris, *The Genesis Flood* (Phillipsburg: P&R Publishing, 1961).

6. Lynn Mitchell and Kirk Blackard, *Reconciling the Bible and Science: A Primer on the Two Books of God* (Charleston: BookSurge Publishing, 2009), 112–13.

7. Carol Hill, Greg Davidson, Tim Helble, and Wayne Ranney, eds., *The Grand Canyon: Monument to an Ancient Earth* (Grand Rapids: Kregel Publications, 2016), 21–29.

8. *Is Genesis History?* https://www.imdb.com/title/tt6360332/; also see clips from the film at https://www.youtube.com/channel/UCzjPwFPxtpZTJ1dq7cAkb3g/featured.

9. The famous quote by Theodosius Dobzhansky, "nothing in biology makes sense except in light of evolution," is the title of his 1973 essay in *The American Biology Teacher*, 35:3 (March 1973), 125–29; https://online.ucpress.edu/abt/article/35/3/125/9833/Nothing-in-Biology-Makes-Sense-except-in-the-Light.

10. *Is Genesis History?* (Del Tackett quote at 1:39:16), https://youtu.be/UM82qxxskZE.

11. Carol Hill et al., *The Grand Canyon: Monument to an Ancient Earth* (Grand Rapids, MI: Kregel Publications, 2016), 127.

12. See beautiful photos of fossilized dunes in Zion National Park: "Sand Dunes to Sandstone," https://www.nps.gov/zion/learn/nature/sand-dunes-sandstone.htm; and in Red Cliff Desert Reserve, Kenyon Virchow, "A Lesson in Geology: Utah's Petrified Sand Dunes," KÜHL, April 10, 2017, https://www.kuhl.com/borninthemountains/utah-petrified-sand-dunes/.

13. Bonneville Salt Flats, Utah.com, https://utah.com/bonneville-salt-flats.

14. Ken Ham, "Billions of Dead Things," https://answersingenesis.org/media/audio/answers-with-ken-ham/volume-137/billions-of-dead-things/.

15. "Catastrophism," *Answers in Genesis*, https://answersingenesis.org/geology/catastrophism/.

16. Henry M. Morris, "Why Christians Should Believe in a Global Flood," Institute for Creation Research, August 1, 1988, https://www.icr.org/article/why-christians-should-believe-global-flood.

17. Donald R. Prothero, *Evolution: What the Fossils Say and Why It Matters* (New York: Columbia University Press, 2017), 66–67.

Chapter 7

1. Kate Carter, "The Curious World of Walter Potter—in Pictures," *The Guardian*, September 13, 2013, https://www.theguardian.com/lifeandstyle/gallery/2013/sep/13/curious-world-walter-potter-pictures-taxidermist-victorian.

2. Read more about Mary Anning and see examples of her spectacular finds: Marie-Claire Eylott, "Mary Anning: The Unsung Hero of Fossil Discovery," Natural History Museum, https://www.nhm.ac.uk/discover/mary-anning-unsung-hero.html.

3. Paul F. Taylor, "How Did the Animals Spread Over the World from Where the Ark Landed?," Answers in Genesis, October 17, 2014, https://answersingenesis.org/animal-behavior/migration/how-did-animals-spread-from-where-ark-landed/.

4. John C. Whitcomb and Henry M. Morris, *The Genesis Flood* (Phillipsburg: P&R Publishing, 1961).

5. Andrew A. Snelling, "Where Are All the Human Fossils?," Answers in Genesis, December 1, 1991, https://answersingenesis.org/fossils/fossil-record/where-are-all-the-human-fossils/; and "Doesn't the Order of Fossils in the Rock Record Favor Long Ages?," Answers in Genesis, February 25, 2014,

https://answersingenesis.org/fossils/fossil-record/doesnt-order-of-fossils-in-rock-favor-long-ages/.

6. Read more about fossils found in Badlands National Park, South Dakota, at the website for the National Park Service, https://www.nps.gov/badl/index.htm.

7. Donald R. Prothero, *Evolution: What the Fossils Say and Why It Matters* (New York: Columbia University Press, 2017), 67–68.

8. Kenneth R. Miller, *Finding Darwin's God* (New York: Cliff Street Books, 1999), 61.

9. It is important to note that the term "kind" is not a term used in modern biology. The concept of "kind" is unique in creationist literature: Jean Lightner, Tom Hennigan, Georgia Purdom, and Bodie Hodge, "Determining the Ark Kinds," Answers in Genesis, August 1, 2014, https://answersingenesis.org/noahs-ark/determining-the-ark-kinds/.

10. Paul F. Taylor, "How Did Animals Spread All Over the World from Where the Ark Landed?," Answers in Genesis, October 18, 2007, https://answersingenesis.org/animal-behavior/migration/how-did-animals-spread-from-where-ark-landed/; Harry F. Sanders III and Troy Lacey, "Floating Log Rafts," Answers in Genesis, August 31, 2019, https://answersingenesis.org/the-flood/global/floating-log-rafts/; Dominic Statham, "Natural Rafts Carried Animals around the Globe," Creation Ministries International, April 2011, https://creation.com/natural-rafts-carried-animals-around-the-globe.

11. Michael Oard, "The Ice Age and the Genesis Flood," Institute for Creation Research, June 1, 1987, https://www.icr.org/article/ice-age-genesis-flood/.

12. Taylor, "How Did Animals Spread?"

13. Taylor, "How Did Animals Spread?"

14. Taylor, "How Did Animals Spread?"

Chapter 8

1. "Pound-stones" are fossilized sea urchins, common in ancient shallow seas. See "Warwickshire Pound-Stones," *Our Warwickshire*, https://www.ourwarwickshire.org.uk/content/article/warwickshire-pound-stones.

2. Read more about William Smith, his faith, and his discovery of faunal succession in Allan Chapman, *Slaying the Dragons: Destroying Myths in the History of Science and Faith* (Oxford: Lion Hudson, 2013), 126.

3. Jerry A. Coyne, *Why Evolution Is True* (New York: Viking, 2009), 53.

4. Donald Prothero, *Evolution: What the Fossils Say and Why It Matters* (New York: Columbia University Press, 2017), 178.

5. See Stephen Meyer, *Darwin's Doubt: The Explosive Origin of Animal Life and the Case for Intelligent Design* (New York: HarperOne, 2013); see also Brian Thomas, "What Were the First Animals Like?," Institute for Creation Research, September 16, 2013, https://www.icr.org/article/what-were-first-animals-like; and Elizabeth Mitchell, "Multicellular Life Evolving in Microfossils, Evolutionists Say," Answers in Genesis, October 23, 2014, https://answersingenesis.org/fossils/transitional-fossils/multicellular-life-evolving-microfossils-evolutionists-say/.

Chapter 9

1. Zack Kopplin, "Showdown Over Science in Texas," *Slate*, September 20, 2013, https://slate.com/technology/2013/09/texas-science-textbooks-creationists-try-to-remove-evolution-from-classrooms.html.

2. David Masci, "For Darwin Day, 6 Facts about the Evolu-

tion Debate," Pew Research Center, February 11, 2019, https://www.pewresearch.org/fact-tank/2019/02/11/darwin-day/.

3. *Is Genesis History?* https://www.imdb.com/title/tt6360332/.

4. This quote is by narrator Del Tackett, in "Q&A with Del Tackett and the Scientists," originally shown in theaters following the feature film. You can watch the Q&A here: https://isgenesishistory.com/qa-with-del-tackett/.

5. Neil Shubin, *Your Inner Fish* (New York: Pantheon, 2008). PBS (2014) created a three-part series based on the book: http://www.pbs.org/your-inner-fish/about/overview/.

6. Donald R. Prothero, *The Story of Life in 25 Fossils* (New York: Columbia University Press, 2015), 111–24.

7. Read more about *Archaeopteryx*: Michael Greshko, "This Famous Dinosaur Could Fly—But Unlike Anything Alive Today," *National Geographic*, March 13, 2018, https://www.nationalgeographic.com/news/2018/03/archaeopteryx-flight-dinosaurs-birds-paleontology-science/.

8. Read more about whale evolution: Kate Wong, "Whence Whales?," *Scientific American*, September 24, 2001, https://www.scientificamerican.com/article/whence-whales/.

9. Frank Turek, in an interview with Cameron Bertuzzi of Capturing Christianity: *Dr. Frank Turek Answers Questions on Evolution, Apologetics, and Memes*, April 20, 2020, https://youtu.be/5JBDTUqOs6w.

10. Neil Shubin, *Some Assembly Required* (New York: Pantheon, 2020), 75–77.

11. Shubin, *Some Assembly Required*, 75–77.

12. Ed Yong, "What a Legless Mouse Tells Us about Snake Evolution," *Atlantic*, October 20, 2016, https://www.theatlantic.com/science/archive/2016/10/what-a-legless-mouse-tells-us-about-snake-evolution/504779/.

Chapter 10

1. Travis M. Andrews, "Nordstrom Selling Jeans Caked in Fake Dirt for Hundreds of Dollars," *Washington Post*, April 26, 2017, https://www.washingtonpost.com/news/morning-mix/wp/2017/04/26/nordstrom-is-selling-jeans-caked-in-fake-dirt-for-hundreds-of-dollars/.

2. Phillip E. Johnson, *Darwin on Trial* (Washington, DC: Regenery Gateway, 1991), 98.

3. Michael J. Behe, *Darwin's Black Box* (New York: The Free Press, 1996).

4. A Rube Goldberg machine consists of a series of steps, each dependent on the next, in order to accomplish some sort of task: Wikipedia, s.v. "Rube Goldberg machine," last modified November 18, 2020 at 22:44 (UTC), https://en.wikipedia.org/wiki/Rube_Goldberg_machine.

5. NOVA produced an award-winning documentary about the Kitzmiller v. Dover trial, *Judgement Day: Intelligent Design on Trial*, https://www.pbs.org/wgbh/nova/video/judgment-day-intelligent-design-on-trial/.

6. Kenneth R. Miller, *Only a Theory* (New York: Viking, 2008), 73.

7. Miller, *Only a Theory*, 178–79.

8. Miller, *Only a Theory*, 114–16.

9. Laurie Goodstein, "Judge Rejects Teaching Intelligent Design," *The New York Times*, December 21, 2005, https://www.nytimes.com/2005/12/21/education/judge-rejects-teaching-intelligent-design.html.

10. Rob Boston, "Missionary Man," *Church & State Magazine*, April 1999, https://www.au.org/church-state/april-1999-church-state/featured/missionary-man.

11. Presentation by Stephen Meyer at Southwestern Assemblies of God University chapel: *Stephen Meyer Explains Neo-Darwinism's False Beliefs-Part 1*, ThoughtHub, SAGU, April 17,

2018, https://www.sagu.edu/thoughthub/thoughthub; also see David Klinghoffer, "Bechly: In the Fossil Record, 'Abrupt Appearances Are the Rule,'" *Evolution News*, February 20, 2018, https://evolutionnews.org/2018/02/bechly-in-the-fossil-record-abrupt-appearances-are-the-rule/.

12. Kenneth R. Miller, *Finding Darwin's God* (New York: Cliff Street Books, 1999), 96–98.

13. Miller, *Finding Darwin's God*, 99.

14. David Klinghoffer, "Q&A with Behe: Your Thoughts on Common Descent?," *Evolution News*, November 27, 2019, https://evolutionnews.org/2019/11/qa-with-behe-your-thoughts-on-common-descent/.

15. Miller, *Only a Theory*, 54–55.

16. K. R. Miller, "Deconstructing Design: A Strategy for Defending Science," Cold Spring Harbor Symposia on Quantitative Biology, 2009.

17. Jerry A. Coyne, *Why Evolution Is True* (New York: Viking, 2009), 139–40.

Chapter 11

1. The Butler Act, State of Tennessee, 1925: http://www.famous-trials.com/scopesmonkey/2128-evolutionstatues.

2. "No Consensus, and Much Confusion, on Evolution and the Origin of Species," *Harris Interactive*, February 18, 2009, https://theharrispoll.com/wp-content/uploads/2017/12/Harris-Interactive-Poll-Research-BBC-Darwin-2009-02.pdf.

3. Irritation + entertainment = irritainment.

4. Wikipedia, s.v. "March of Progress," last modified November 17, 2020 at 15:52 (UTC), https://en.wikipedia.org/wiki/March_of_Progress.

5. Maya Wei-Haas, "Ancient Girl's Parents Were Two Different Human Species," *National Geographic*, August 22, 2018,

https://www.nationalgeographic.com/science/2018/08/news-denisovan-neanderthal-hominin-hybrid-ancient-human/.

6. Peter Andres and Christopher Stringer, "The Primates Progress," in Steve Jay Gould, ed., *The Book of Life* (New York: W. W. Norton & Company, 2001), 249–50.

7. Maya Wei-Haas, "You May Have More Neanderthal DNA Than You Think," *National Geographic*, January 30, 2020, https://www.nationalgeographic.com/science/2020/01/more-neanderthal-dna-than-you-think/.

8. Michael Price, "Africans Carry Surprising Amount of Neanderthal DNA," *Science*, January 30, 2020, https://www.sciencemag.org/news/2020/01/africans-carry-surprising-amount-neanderthal-dna.

9. Matthew Warren, "Biggest Denisovan Fossil Yet Spills Ancient Human's Secrets," *Nature*, May 1, 2019, https://www.nature.com/articles/d41586-019-01395-0.

10. Bernard Wood, "Welcome to the Family," *Scientific American* 311, no. 3 (Sept. 2014), 40–47.

11. "What Makes Us Special," *Scientific American* 311, no. 3 (Sept. 2014), 60–61.

12. David Menton, "Did Humans Really Evolve from Apelike Creatures?," Answers in Genesis, February 25, 2010, https://answersingenesis.org/human-evolution/ape-man/did-humans-really-evolve-from-apelike-creatures/.

13. A PBS interview with Francis Collins on the program *The Question of God*: https://www.pbs.org/wgbh/questionofgod/voices/collins.html.

14. "The Human Genome Project," National Human Genome Research Institute, https://www.genome.gov/human-genome-project.

15. Francis S. Collins, "Remarks at the Press Conference Announcing Sequencing and Analysis of the Human Genome," National Human Genome Research Institute, https://

www.genome.gov/10001379/february-2001-working-draft-of-human-genome-director-collins.

16. Francis S. Collins, *The Language of God* (New York: Free Press, 2006), 137–38.

17. Kenneth R. Miller, *Only a Theory* (New York: Viking, 2008), 105–7.

18. See Jean Lightner, "Chromosome Tales and the Importance of a Biblical Worldview," Answers in Genesis, June 18, 2014, https://answersingenesis.org/genetics/dna-similarities/chromosome-tales-and-importance-biblical-worldview/; and Jeffrey P. Tomkins, "New Research Debunks Human Chromosome Fusion," Institute for Creation Research, November 27, 2013, https://www.icr.org/article/new-research-debunks-human-chromosome/.

19. See Nathaniel T. Jeanson, "An Update on Chromosome 2 'Fusion,'" Institute for Creation Research, August 30, 2013, https://www.icr.org/article/update-chromosome-2-fusion; and "Human-Chimp Genetic Similarity: Refuting the Appeal to Human Genetics," Institute for Creation Research, July 29, 2011, https://www.icr.org/article/chimp-similarity-refuting-appeal-human.

20. Elie Dolgin, "Human Mutation Rate Revealed," *Nature*, August 27, 2009, https://www.nature.com/articles/news.2009.864.

21. Dennis R. Venema and Scot McKnight, *Adam and the Genome: Reading Scripture after Genetic Science* (Grand Rapids: Brazos Press, 2017), 43–55.

22. Nathaniel Jeanson and Jason Lisle, "On the Origin of Eukaryotic Species' Genotypic and Phenotypic Diversity," Answers in Genesis, April 20, 2016, https://answersingenesis.org/natural-selection/speciation/on-the-origin-of-eukaryotic-species-genotypic-and-phenotypic-diversity/.

23. Read more about interpretations of the Adam and Eve

story: Dennis Venema and Scot McKnight, *Adam and the Genome*; Peter Enns, *The Evolution of Adam: What the Bible Does and Doesn't Say about Human Origins* (Grand Rapids, MI: Brazos Press, 2012); John H. Walton, *The Lost World of Genesis One: Ancient Cosmology and the Origins Debate* (Downers Grove, IL: IVP Academic, 2009); Stephen C. Barton and David Wilkinson, eds., *Reading Genesis after Darwin* (Oxford: Oxford University Press, 2009).

Chapter 12

1. Michael Hoskin, ed., *The Cambridge Illustrated History of Astronomy* (Cambridge: Cambridge University Press, 1997), 130.

2. Psalm 96:10 (KJV); Psalm 104:5 (KJV); Joshua 10:13 (KJV).

3. Hoskin, *The Cambridge Illustrated History of Astronomy*, 130; see also Allan Chapman, *Stargazers: Copernicus, Galileo, the Telescope, and the Church* (Oxford: Lion Books, 2014).

4. Job 38:22–25; Deuteronomy 28:12; Job 37:6; Psalm 135:7; Job 37:11–12.

5. Psalm 139:13.

6. Hoskin, *The Cambridge Illustrated History of Astronomy*, 162.

7. Lucas Brouwers, "Did Life Evolve in a 'Warm Little Pond'?," Thoughtomics, *Scientific American*, February 16, 2012, https://blogs.scientificamerican.com/thoughtomics/did-life-evolve-in-a-warm-little-pond/.

8. Deborah Haarsma, "Learning to Praise God for His Work in Evolution," in *How I Changed My Mind about Evolution*, ed. Kathryn Applegate and J. B. Stump (Downers Grove: InterVarsity Press, 2016).

9. "How Great Thou Art," words by Stuart K. Hine.

10. "Atheism Doubles among Generation Z," Barna, Jan-

uary 24, 2018, https://www.barna.com/research/atheism-doubles-among-generation-z/.

11. Two excellent reads on current church trends are David Kinnaman, *You Lost Me: Why Young Christians Are Leaving Church . . . and Rethinking Faith* (Grand Rapids: Baker Books, 2011) and George Barna and David Kinnaman, (eds.), *Churchless* (Tyndale, 2014).

12. "Six Reasons Young Christians Leave Church," Barna, September 27, 2011, https://www.barna.com/research/six-reasons-young-christians-leave-church/#.V1IFAI-cGUk.

13. *Bill Nye Debates Ken Ham*, https://www.youtube.com/watch?v=z6kgvhG3AkI.

14. Andrew Root, David Wood, Tony Jones, "Youth Ministry and Science," a Templeton Planning Grant, January 2015.

15. Kathryn Applegate and J. B. Stump, eds., *How I Changed My Mind about Evolution*.

16. See for example Henry M. Morris, "The Uncertain Speed of Light," Institute for Creation Research, June 1, 2003, https://www.icr.org/article/uncertain-speed-light/; and Eric Hovind, "How Could Light Have Travelled Millions of Years?," Creation Today, https://creationtoday.org/how-could-light-travel-millions-of-years/.

17. David Masci, "For Darwin Day, 6 Facts about the Evolution Debate, Pew Research Center, February 11, 2019, https://www.pewresearch.org/fact-tank/2019/02/11/darwin-day/.

18. Numbers 23:19.

19. Sean Flynt, "A House Is Not a Home: Science, Genesis, Tell Different Stories, Walton Says," Samford University, April 12, 2016, https://www.samford.edu/news/2016/04/A-House-is-Not-a-Home-Science-Genesis-Tell-Different-Stories-Walton-Says.

20. See Peter Enns's *The Evolution of Adam: What the Bible Does and Doesn't Say about Human Origins* (Grand Rap-

ids, MI: Brazos Press, 2012) for understanding genre and the ancient voice of Genesis; see Scot McKnight (with Dennis R. Venema) in *Adam and the Genome* (Grand Rapids: Brazos Press, 2017) for an overview of theological interpretations of the creation story.

21. An interview with Francis Collins: "God Is Not Threatened by Our Scientific Adventures" on Beliefnet, 2006, https://www.beliefnet.com/news/science-religion/2006/08/god-is-not-threatened-by-our-scientific-adventures.aspx.

22. Acts 17:28.

Index